Nicholas J. Conard · Henning Hassmann · Kurt Felix Hillgruber
Jordi Serangeli · Thomas Terberger (eds)

The *Homotherium* Finds from Schöningen 13II-4
Man and Big Cats of the Ice Age

FORSCHUNGEN ZUR URGESCHICHTE AUS DEM TAGEBAU VON SCHÖNINGEN

Band 4

SENCKENBERG
world of biodiversity

EBERHARD KARLS
UNIVERSITÄT
TÜBINGEN

Niedersächsisches Landesamt
für Denkmalpflege

Römisch-Germanisches
Zentralmuseum
Leibniz-Forschungsinstitut
für Archäologie

RGZM

DEUTSCHES
ARCHÄOLOGISCHES INSTITUT

Römisch-Germanisches Zentralmuseum
Leibniz-Forschungsinstitut für Archäologie

Senckenberg Centre for Human Evolution
and Palaeoenvironment
an der Universität Tübingen

Römisch-Germanische Kommission
des Deutschen Archäologischen Instituts

Niedersächsisches Landesamt für Denkmalpflege
Hannover

Nicholas J. Conard · Henning Hassmann · Kurt Felix Hillgruber
Jordi Serangeli · Thomas Terberger (eds)

THE *HOMOTHERIUM* FINDS FROM SCHÖNINGEN 13II-4

MAN AND BIG CATS OF THE ICE AGE

Contributions of the scientific workshop
at the paläon (Schöningen) from 05.06. to 07.06.2015

Verlag des Römisch-Germanischen Zentralmuseums Mainz 2022

Gefördert durch

Niedersächsisches Landesamt
für Denkmalpflege

FÖRDERVEREIN SCHÖNINGER SPEERE
ERBE DER MENSCHHEIT E.V.

Redaktion: Kurt Felix Hillgruber (Konstanz),
Ivo Verheyen (Schöningen), Marie Reiter (RGZM)
Satz: Claudia Nickel (RGZM)
Umschlaggestaltung: Claudia Nickel (RGZM)
unter Verwendung eines Bildes von Mauricio Antón

Bibliografische Information
der Deutschen Nationalbibliothek

Die Deutsche Nationalbibliothek verzeichnet diese Publikation in der Deutschen Nationalbibliografie: Detaillierte bibliografische Daten sind im Internet über **http://dnb.d-nb.de** abrufbar.

ISBN 978-3-88467-339-3

Diese Publikation ist auf http://www.propylaeum.de dauerhaft frei verfügbar (Open Access).
DOI: https://doi.org/10.11588/propylaeum.1006
eISBN: 978-3-96929-136-8 (PDF)
URN: urn:nbn:de:bsz:16-propylaeum-ebook-1006-7

Propylaeum
FACHINFORMATIONSDIENST
ALTERTUMSWISSENSCHAFTEN

Druck: Memminger MedienCentrum Druckerei und Verlags-AG
Printed in Germany.

CONTENTS

Nicholas J. Conard · Henning Hassmann · Kurt Felix Hillgruber
Jordi Serangeli · Thomas Terberger
Foreword / Vorwort VII

Homotherium

Ivo Verheijen · Jordi Serangeli · André Ramcharan · Britt Starkovich · Thijs van Kolfschoten
The Remains of the European Sabre-toothed Cat (*Homotherium latidens*)
Found in the Spear Horizon at Schöningen (Germany) 3

Mauricio Antón
Behaviour of *Homotherium* in the Light of Modern African Big Cats 19

Dick Mol · Bram W. Langeveld
On Remains of *Homotherium* from the Bottom of the North Sea
between the British Isles and the Netherlands and their Stratigraphical Age 35

Ross Barnett
A Very British Sabretooth: *Homotherium* in the UK 53

Extinction

Ulrich Joger
Ice Age Fauna in the Braunschweig Region, with a Focus on the Carnivores 69

Aritza Villaluenga
Biogeographic Analysis of Upper Pleistocene Felid (*Felis*, *Lynx* and *Panthera*) Remains
in Archaeological Sites on the Iberian Peninsula 79

Anthony J. Stuart
The Extinction of Large Carnivores in Northern Eurasia 99

Art

Kurt Felix Hillgruber · Thomas Terberger
Big Cats and Humans – Some Remarks on the Interaction with Big Cats in Prehistoric Art 109

Nicolas Zenzen
The Lion in Ancient Art – The Story of a Striking Success 119

Modern Cats – Carnivores

Ulrich Schmölcke · Elena A. Nikulina
Humans and their Relationship to Large Carnivores in Central Europe
from Foragers to Modern Times: A Survey 129

Clara Stefen
The European Wildcat *Felis silvestris* – A Brief Review of Status and Research in Germany 151

Fritz Jantschke
Tiger and Man – The Difficult Co-Existence of Two Top Predators in India 165

Klaus Müller-Schilling
Breeding Conservation of Endangered Big Cats – The Example of Barbary Lions 187

List of Contributors 205

FOREWORD / VORWORT

The discovery of the remains of European sabre-toothed cats (*Homotherium latidens*), also colloquially known as sabre-toothed tigers, during excavations in the Schöningen open-cast mine from 2012 to 2014, was generally viewed as a sensation. Additional important finds from a place where many important discoveries have already been made. The present book is the fourth monograph in the series »Forschungen zur Urgeschichte aus dem Tagebau von Schöningen« and the second volume based on a colloquium. Due to these exciting new findings and in preparation for the exhibition »The Ice Age Huntress«, a two-day scientific workshop was held in 2015 with a focus on sabre-toothed cats and other cats of prey. Oral workshop contributions were summarized by the authors and are now combined into this volume. After three previous volumes were devoted to the chronology, geology and botanical finds of the Schöningen excavations, this volume contains a wide range of zoological topics, thereby highlighting the important contributions of the Schöningen excavations to the study of the Middle Pleistocene in Europe. The present volume provides an overview of a topic crucial to human development by focusing on our evolutionary coexistence with large predatory cats.

In 2012 two of the European sabre-toothed cats were discovered in the same find layer where two decades earlier the world-famous Schöningen spears were excavated. The third individual was discovered in a slightly deeper layer. These finds once again focus on the evolution of our ancestors in the light of their coexistence with other large predatory fauna. The discovery of the *Homotherium* remains demonstrates that 300,000 years ago these large and powerful feline predators cohabited at the Schöningen lake shore together with our human ancestors. One long bone of one of these sabre-toothed cats was even manipulated and used as a tool by *Homo heidelbergensis*, representing the only tool worldwide made from the bones of a sabre-toothed cat.

These finds are a testimony to our long, shared relationship with large predators, a relationship determined by competition, struggle for survival and innovation. Firstly, these bone finds demonstrate that members of the genus *Homotherium* were still native to Europe during the Middle Pleistocene. Thus, their extinction in Europe must be revaluated. On the other hand, Schöningen also provides evidence for the direct contact between these impressive predators and our human ancestors. The wooden weapons of *Homo heidelbergensis* from Schöningen are probably a further evidence of this coexistence, which undoubtedly influenced our cultural evolution and our struggle to survive against predatory megafauna, but also might explain that today almost all big cats are threatened with extinction.

Starting with the Schöningen finds, this volume presents the fossil evidence of the European sabre-toothed cat, spanning from the type fossil from Kents Cavern to the fossil remains collected without stratigraphic context from the bottom of the North Sea. The latter, as well as the finds from Schöningen clearly demonstrate that there are still many open questions regarding population dynamics and extinction events during the Pleistocene in Europe. The discovery of sabre-toothed cat remains around the world with their respective contexts allows us to better understand the behaviour and way of life of these big cats. In this context, the actualistic comparison with today's predatory cats is here of particular importance. In addition, the aim of the workshop was also to discuss the relationship between big cats and our human ancestors. This is apparent by the deep respect we humans still feel towards these large feline predators, even though over time we have increased our position in the food chain. This is particularly demonstrated by representations of big cats in art, from the Palaeolithic onwards. The special relationship between humans and big cats is still evident today, even though we also have to acknowledge that we are causally responsible for the extinction

of these fascinating animals, while at the same time endeavouring to safeguard habitats for lions, tigers and other wild cats and securing their future existence through breeding and species protection programmes.
For their efforts in preparing the texts, illustrations and plates for printing we would like to extend our gratitude to Claudia Nickel, Martin Schönfelder and Ivo Verheijen.
All volumes of this series are available both in print and in digital form from the publisher the Römisch-Germanisches Zentralmuseum in cooperation with the library of the Heidelberg University (Propylaeum Open Access publication platform for scientific ebooks on Classical Studies) and are therefore a contribution to the ongoing efforts of the scientific community to make research results freely accessible to various target groups via open access publications.
Special thanks go to the »Förderverein Schöninger Speere – Erbe der Menschheit e. V.«, which made the above-mentioned large sabre-toothed cat exhibition possible and generously supported the printing of this volume.

Die Entdeckung von Überresten von drei Europäischen Säbelzahnkatzen (*Homotherium latidens*), auch umgangssprachlich Säbelzahntiger genannt, bei den Ausgrabungen im Schöninger Tagebau in den Jahren 2012 bis 2014 war eine Sensation. Weitere einschlägige Funde in einer Fundstelle, die bereits viele bedeutende Funde und Befunde preisgegeben hat.
Das vorliegende Buch ist die vierte Monographie in der Reihe »Forschungen zur Urgeschichte aus dem Tagebau von Schöningen« und der zweite Band, der auf einem Kolloquium basiert. Im Jahr 2015 wurde, aufgrund dieser neuen Funde und in Vorbereitung der Sonderausstellung »Die Eiszeitjägerin«, ein zweitägiger wissenschaftlicher Workshop zum Thema Säbelzahnkatzen und Raubkatzen abgehalten. Dankenswerterweise waren die Vortragenden bereit, ihre Beiträge in Form dieses Bandes zu veröffentlichen. Nachdem die drei vorangehenden Bände der Reihe der Chronologie, Geologie und den botanischen Funden gewidmet waren, wird ein zoologisches Thema nun die große Spannbreite an Themen aufzeigen, in der die Ausgrabungen in Schöningen wichtige Beiträge zur Erforschung des Mittelpleistozäns leisten. Der vorliegende Band gibt einen Überblick über ein für die Entwicklung des Menschen entscheidendes Thema – unser Zusammenleben mit großen Raubkatzen.
Zwei der Überreste der Europäischen Säbelzahnkatzen wurden 2012 in derselben Fundschicht entdeckt, in der zwei Jahrzehnte zuvor die weltberühmten Schöninger Speere gefunden wurden. Das dritte Individuum lag etwas tiefer. Diese Funde werfen erneut ein Schlaglicht auf die Entwicklung unser Vorfahren. Diese großen und kräftigen Raubtiere lebten zeitgleich mit den damaligen archaischen Menschen vor 300 000 Jahren am Ufer des Schöninger Sees. Ein Langknochen einer dieser Säbelzahnkatzen wurde sogar vom *Homo heidelbergensis* manipuliert und als Gerät verwendet – weltweit das einzige Werkzeug aus dem Knochen eines solchen Tieres.
Diese Funde sind ein beredtes Zeugnis unserer langen, gemeinsamen Beziehung zu großen Raubtieren, die von Konkurrenz, Überlebenskampf und Innovationen geprägt ist. Zum einen zeigen diese Knochenfunde von Säbelzahnkatzen, dass *Homotherium* noch während des Mittelpleistozäns in Europa heimisch war. Damit ist ihr Aussterben in Europa neu zu datieren. Zum anderen liefert Schöningen den Beleg für das direkte Nebeneinander dieser beeindruckenden Raubtiere und des Menschen. Die hölzernen Waffen des *Homo heidelbergensis* von Schöningen sind auch eine Antwort auf diesem Zusammenleben, welches zweifelsohne unsere kulturelle Evolution beeinflusste und dazu führte, dass heute fast alle Großkatzen vom Aussterben bedroht sind.

Von den Schöninger Funden ausgehend präsentiert dieser Band die fossilen Belege der europäischen Säbelzahnkatze in Europa, vom Typfossil aus der Kents Cavern bis zu den höchst interessanten und viele Fragen aufwerfenden Knochenfunden ohne stratigraphischen Kontext aus der Nordsee. Letztere und die Funde aus Schöningen bezeugen, dass wir bezüglich Populationsdynamik und Aussterbeereignisse im Pleistozän noch immer viel zu erforschen haben. Weltweite Funde von Säbelzahnkatzen und ihre jeweiligen Fundumstände lassen uns in Ansätzen das Verhalten und Lebensweise dieser Großkatzen besser verstehen, wobei insbesondere der aktualistische Vergleich zu heutigen Raubkatzen von Bedeutung ist. Ziel des Workshops war es aber auch, das Verhältnis von Großkatzen und uns Menschen in der Geschichte zu diskutieren. Auch wenn wir uns in der Nahrungskette stetig weiterentwickelt haben, so tragen wir Menschen doch in unserem Inneren auch heute noch einen tiefen Respekt vor den großen Raubtieren in uns. Davon zeugen Beiträge, die sich mit Darstellungen von Großkatzen in der Kunst beschäftigen, schon vom Paläolithikum an. Die besondere Beziehung von uns Menschen zu Raubkatzen wird aber heute noch augenscheinlich, wenn wir zum einen erkennen müssen, dass wir ursächlich für das Aussterben dieser faszinierenden Tiere verantwortlich sind, zum anderen aber die Bemühungen anerkennen müssen, mit Zuchtprogrammen und Artenschutz auch in der Zukunft einen Lebensraum für Löwen, Tiger und andere Wildkatzen auf unserer Welt zu garantieren.

Für ihren Einsatz zur druckfertigen Aufarbeitung des Textes sowie der Abbildungen und Tafeln möchten wir Claudia Nickel, Martin Schönfelder und Ivo Verheijen ganz herzlich danken.

Die Bände dieser Reihe sind sowohl als Printausgabe als auch in digitaler Form über den Verlag des Römisch-Germanischen Zentralmuseums in Kooperation mit der Universitätsbibliothek Heidelberg (Propylaeum Fachinformationsdienst Altertumswissenschaft) erhältlich und entsprechen somit dem Bestreben der Scientific Community, Forschungsergebnisse über Open Access-Veröffentlichungen verschiedenen Zielgruppen frei zugänglich zu machen.

Besonderer Dank gebührt dem »Förderverein Schöninger Speere – Erbe der Menschheit e. V.«, der die o. g. große Säbelzahnkatzenausstellung ermöglicht und den Druck dieses Bandes großzügig unterstützt hat.

Nicholas J. Conard, Henning Hassmann, Kurt Felix Hillgruber,
Jordi Serangeli, Thomas Terberger
im November 2021

HOMOTHERIUM

IVO VERHEIJEN · JORDI SERANGELI · ANDRÉ RAMCHARAN
BRITT STARKOVICH · THIJS VAN KOLFSCHOTEN

THE REMAINS OF THE EUROPEAN SABRE-TOOTHED CAT (*HOMOTHERIUM LATIDENS*) FOUND IN THE SPEAR HORIZON AT SCHÖNINGEN (GERMANY)

In the last 25 years, over 20 000 fossil animal remains were recovered at the Palaeolithic sites of Schöningen (Lkr. Helmstedt/D) and analysed by various researchers (van Asperen 2004; Voormolen 2008; van Kolfschoten 2014; van Kolfschoten/Buhrs/Verheijen 2015; van Kolfschoten et al. 2015; Starkovich/Conard 2015). The mammalian faunal assemblages include not only species that are common during the late Middle Pleistocene, but also species that are rare, for example the giant beaver (*Trogontherium cuvieri*) (Heinrich/van Kolfschoten 2007) and the water buffalo (*Bubalus murrensis*) (Serangeli et al. 2015a). Here we report the remains of another rare species, the European sabre-toothed cat, *Homotherium latidens* (Owen, 1846), found at the Schö 13II site (**figs 1-2**). In the preliminary description (Serangeli et al. 2015b), two individuals of *Homotherium* were described from the Spear Horizon site (layer Schö 13II-4a-4b/c) in Schöningen. A third individual, represented by eleven cranial fragments, was recently discovered in layer Schö 13II-4e, stratigraphically situated over 1 m below the level of the other *Homotherium* finds.
With an approximate age of 300 000-320 000, the three individuals from Schöningen are among the youngest remains of *Homotherium latidens* with clear stratigraphic provenance and in an archaeological context. The context, together with hominin modifications present on one of the humeri, confirms that Middle Pleistocene hominins shared their environment with the sabre-toothed cat. The presence of sabre-toothed cats certainly had an impact on Lower Palaeolithic hominin daily subsistence and behaviour.

HOMOTHERIUM LATIDENS

Homotherium was a wide-spread member of the Pliocene-Early Pleistocene carnivore guild of Africa, Eurasia and the Americas; the genus was only absent in Australia/Oceania and on the Antarctic continent (Rincón/Prevosti/Parra 2011). Its wide distribution indicates adaptation to various climatic zones and environments. However, several authors consider open landscapes as the typical habitat of members of the genus *Homotherium* (Antón 2013; Naples/Martin/Bibiarz 2011).
Some authors (e. g. Ficcarelli 1979; Turner/Antón 1997; Sotnikova/Titov 2009) distinguish two chronospecies from the Eurasian fossil record: *Homotherium crenatidens*, with a Villafranchian age (± 3.0-1.0 Ma), and *Homotherium latidens* with a post-Villafranchian age (< 1.0 Ma). The differences between these species are, however, so minimal that Antón et al. (2014) assume that all Pleistocene *Homotherium* remains from Europe should be referred to as one species, *Homotherium latidens*. The *Homotherium* assemblage from Incarcal (prov. Girona/E; Galobart et al. 2003) supports this assertion. The observed, large intraspecific variation in size and mandibular morphology is interpreted as sexual dimorphism.

In: Nicholas J. Conard · Henning Hassmann · Kurt Felix Hillgruber · Jordi Serangeli · Thomas Terberger (eds), The *Homotherium* Finds from Schöningen 13II-4. Man and Big Cats of the Ice Age. Forschungen zur Urgeschichte aus dem Tagebau von Schöningen 4 (Mainz 2022). DOI: https://doi.org/10.11588/propylaeum.1006.c13518

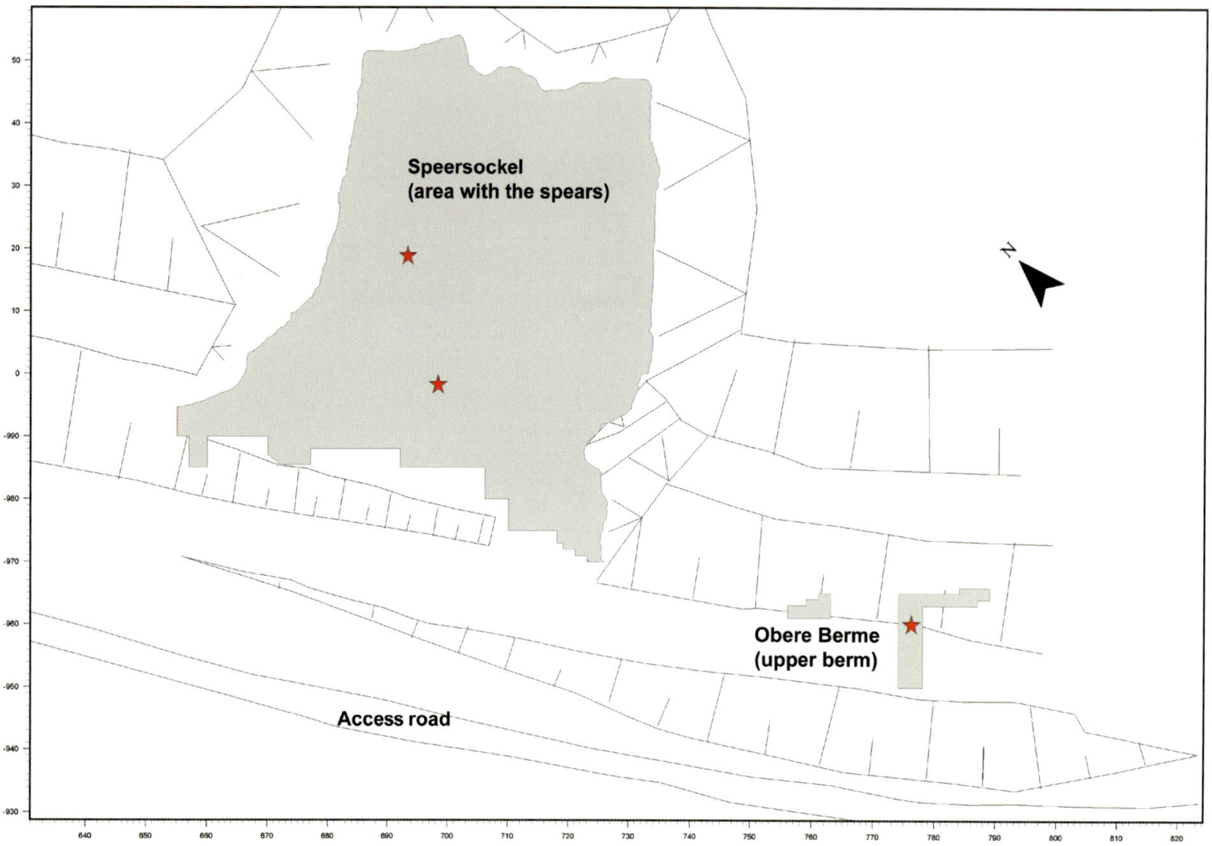

Fig. 1 Schöningen (Lkr. Helmstedt/D): Map of the excavated areas in the Spear Horizon, the »Speersockel« (the area with the spears; excavated from 1994 until 2012), and the new area »Obere Berme« (upper berm; excavated since 2011 until today). The stars indicate the position of the remains from the three *Homotherium*. – (Illustration J. Lehmann).

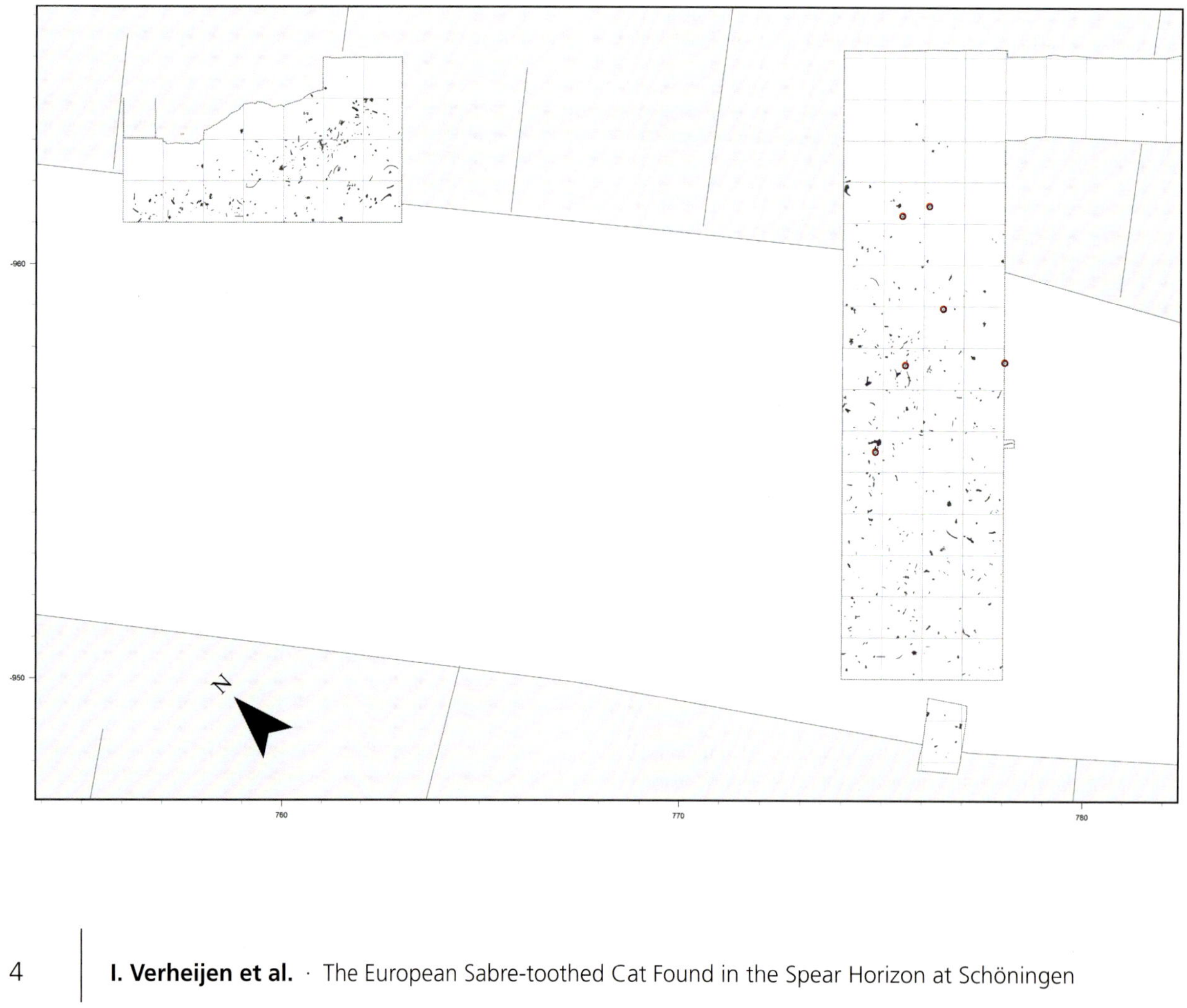

	description	individual	refit
ID 26509	I1 sup. sin.	1	
ID 26392	canine sup. medial fragment	1	ID 27746 ID 27872
ID 27872	canine sup. fragment of posterior cutting edge	1	ID 26392 ID 27746
ID 27746	canine sup. fragment of the posterior cutting edge	1	ID 27872 ID 26392
ID 27876	canine sup. (?) enamel fragment of the medial/lateral side	1	
ID 26238	P4 sup. (posterior half)	1	ID 26470
ID 26470	P4 sup. (anterior half)	1	ID 26238
ID 26367	M1 sup. sin./dex.	1	
ID 26229	i3 inf. dex.	1	
ID 27749	p3 inf. sin.	1	
ID 26549	m1 inf. dex.	1	
ID 26221	costa dex.	1	
ID 26207	scapula dex.	1	
ID 26234	humerus dex.	1	
ID 25885	humerus dex.	2	
ID 27378	right temporal/parietal	3	ID 27379, ID 27380, ID 27381, ID 27382, ID 27384
ID 27379	basioccipital and fragment of the ventral bulla tympanica	3	ID 27378, ID 27380, ID 27381, ID 27382, ID 27384
ID 27380	left frontal	3	ID 27378, ID 27379, ID 27381, ID 27382, ID 27384
ID 27381	occipital, fragment of the left mastoid and three fragments of the left glenoid fossa	3	ID 27378, ID 27379, ID 27380, ID 27382, ID 27384
ID 27382	medial part of the right and left parietal, three fragments of the right frontal bone	3	ID 27378, ID 27379, ID 27380, ID 27381, ID 27384
ID 27384	left temporal/parietal	3	ID 27378, ID 27379, ID 27380, ID 27381, ID 27382

Tab. 1 Overview of *Homotherium* finds from Schöningen, organized per individual. Anatomical refits are indicated in the last column.

Homotherium was, on average, about the size of a modern lion or tiger, with a withers-height of 0.9-1.1 m, a total length of 1.5-2 m and a weight of up to 200 kg (Turner/Antón 1997; Hemmer 2001; Antón/Galobart/Turner 2005; Antón 2013). Its long, powerful neck, long front legs and relatively short hind legs resulted in a stance that differs from other big cats and is more similar to that of a hyena or a bear (Hemmer 2001). The long sabre-like canines, together with the serrated incisors, canines and carnassials are the most distinctive characteristics of *Homotherium*. It is much more difficult to define specific features on postcranial elements to differentiate *Homotherium* from other large felids. This aspect, together with the fact that teeth are generally more resistant to natural weathering compared to bones, might explain why a disproportionate number of remains attributed to *Homotherium* are teeth.

Fig. 2 Schöningen (Lkr. Helmstedt/D): Map of the until July 2014 excavated Spear Horizon »Obere Berme«. The red circles indicate the position of the teeth. From top to bottom: ID 26470, ID 26238, ID 26229, ID 26549, ID 29392 and ID 26509. – (Illustration J. Lehmann).

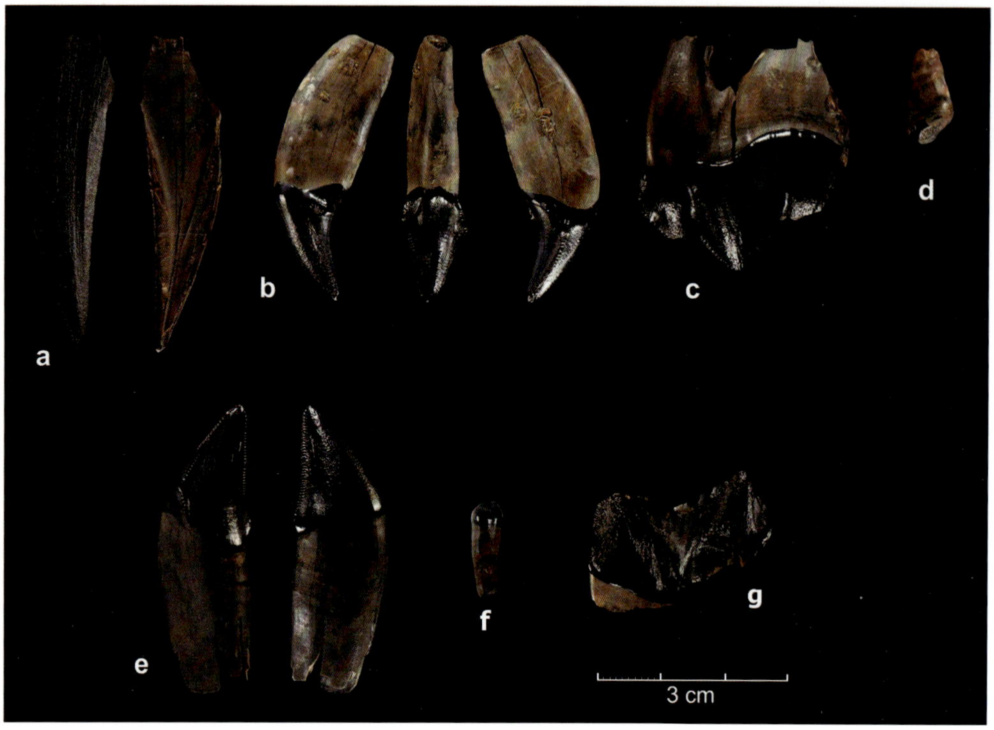

Fig. 3 Schöningen (Lkr. Helmstedt/D): *Homotherium latidens,* dentition: **a** fragment of an upper canine lateral and medial view (ID 26392). – **b** upper I1 sin. (ID 26509), posterior view and side views. – **c** upper P4 sin. (ID 26470, ID 26238), labial view. – **d** upper M1 (ID 26367), posterior view. – **e** lower i3 dex. (ID 26229), side views. – **f** lower p3 sin. (ID 27749), lingual view. – **g** lower m1 dex. (ID 26549), lingual view. – (Photos V. Minkus / I. Verheijen).

DESCRIPTION OF THE SCHÖNINGEN *HOMOTHERIUM* REMAINS

Here we present the remains of three different sabre-toothed cat individuals (**tabs 1-2**). The three individuals have a nickname to honor the person who discovered and excavated the remains.

Individual 1 (Martin, after Martin Kursch)

The seven teeth from the »Obere Berme« (a continuation of the Spear Horizon located to the south of the main concentration; **fig. 3**) are from a subadult *Homotherium* individual, a conclusion that is based on the extremely well-preserved serration of the teeth, with only minor signs of wear, and the open root of the incisors.

The partial denture, discovered in the »Obere Berme« area, is comprised of: 1) a fragmented upper canine, consisting of a medial part (ID 26392) and two small fragments of the posterior edge (ID 27746, ID 27872); 2) a left first upper incisor (ID 26509) with an open root; 3) a fragmented left fourth upper premolar (ID 26470, ID 26238); 4) an upper first molar (ID 26367); 5) a right third lower incisor (ID 26229); 6) a complete left lower p3 (ID 27749); 7) a partly preserved, fragmented first right lower molar (ID 26549). The teeth display a black, shiny patina, consistent with other finds from this horizon. The enamel is crenulated, with a smooth cingulum at the base. Most dental elements (with the exception of the upper M1 and the lower p3) have serration on the crown, with a frequency of 21 serrations per centimeter, observed on the first upper incisor. Several postcranial elements were found in association with the teeth. Their morphology is indicative

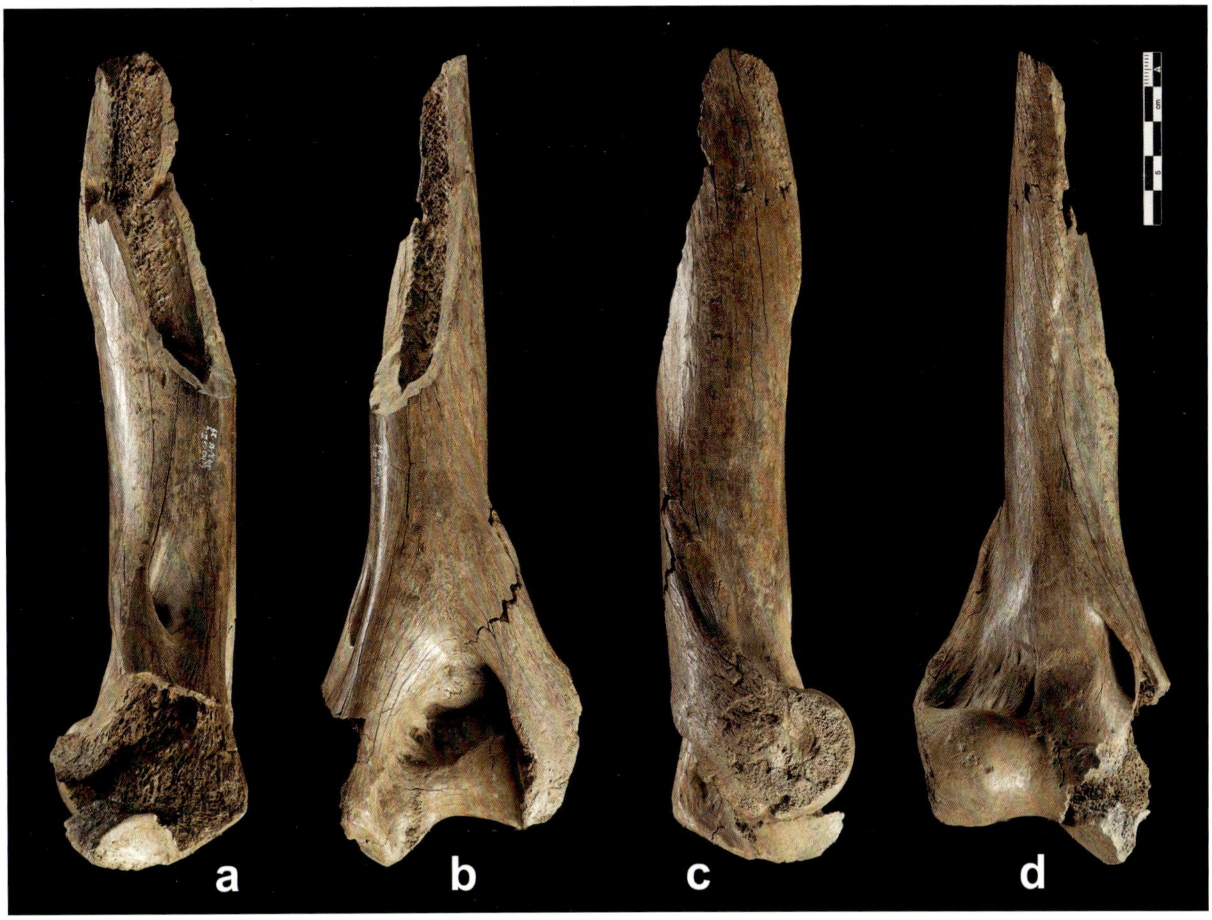

Fig. 4 Schöningen (Lkr. Helmstedt/D): *Homotherium latidens*, distal part of a right humerus: **a** medial view. – **b** posterior view. – **c** lateral view. – **d** anterior view (ID 25885). – (Photo V. Minkus).

of a felid and their bone quality corresponds well with that of a young individual. Due to their fragmentary nature, it is not possible to assign these to a specific species.
Several morphological features of the teeth, such as the basal tubercles on the incisors (Sotnikova/Tivov 2009), and in particular the presence of the serrations, led to the identification of these teeth as belonging to *Homotherium latidens*. In addition, the anterior edge of the parastyle of the upper fourth premolar displays the presence of a preparastyle. This characteristic was formerly associated with *Homotherium moravicum* (Sardella/Iurino 2012), which is now considered to be synonymous with *H. latidens* (Sotnikova/Tivov 2009).

Individual 2 (Wolfgang, after Wolfgang Berkemer)

The second individual is represented by a right humerus (ID 25885), excavated from the »Speersockel« (the area where the wooden spears where discovered; **fig. 4**), located at a distance of 100 m from the aforementioned finds from the »Obere Berme«. The proximal half of the bone is missing, and the distal ephiphysis shows some damage as well. Therefore, multiple characteristics that can be used for a species identification are missing. One characteristic that can be used to differentiate *Homotherium* from other large felids is the shape and position of the supracondyloid foramen (**fig. 4**). In *Homotherium*, this foramen has an oval

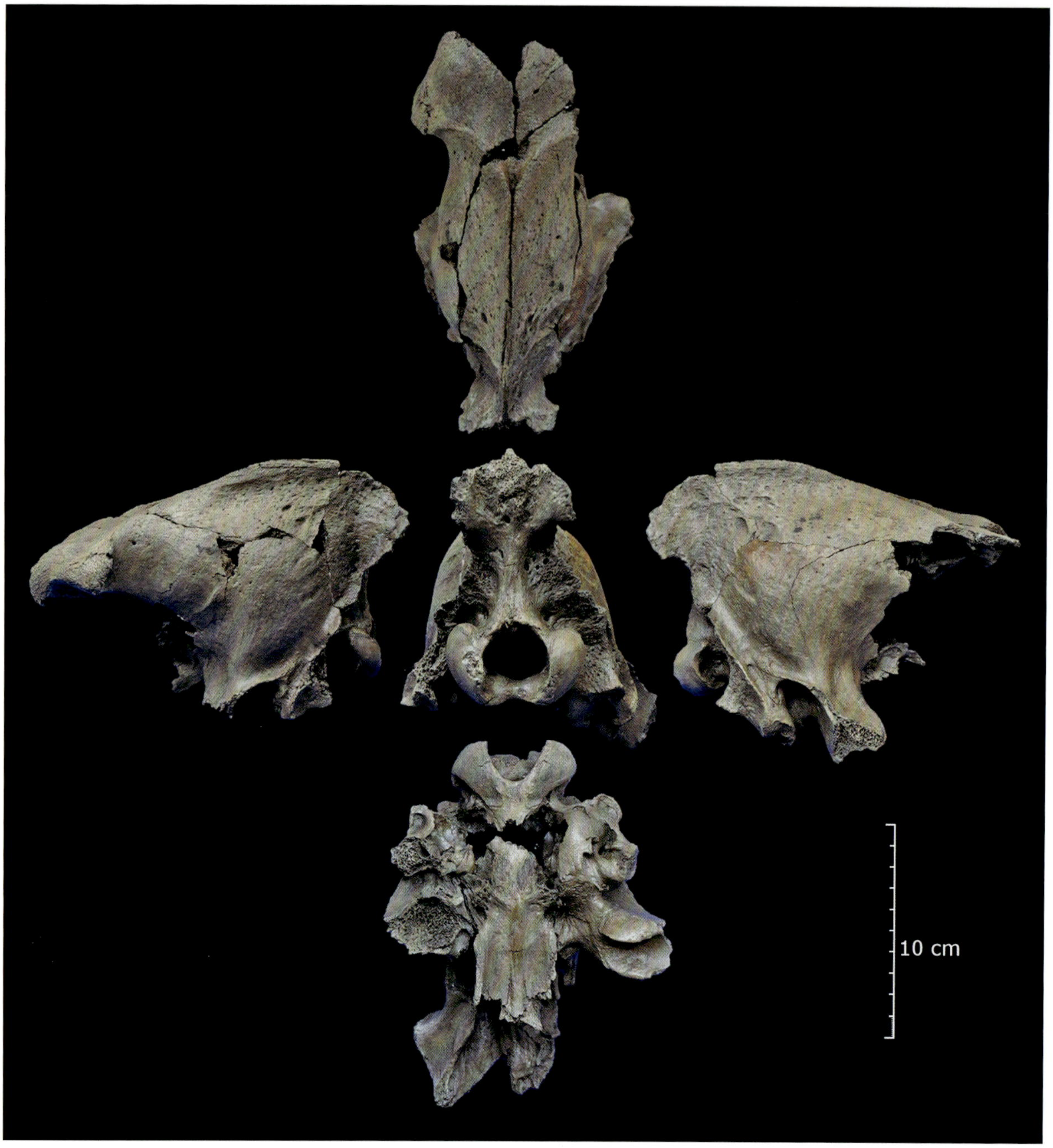

Fig. 5 Schöningen (Lkr. Helmstedt/D): Neurocranium of *Homotherium* individual 3 (Jörg) from layer Schö 13II-4e, after refitting. From top to bottom, left to right: dorsal view, lateral view (left side), posterior view, lateral view (right side) and ventral view. – (Photos I. Verheijen).

shape, whereas in lions it is more rounded. Besides that, in lions it is positioned less medially and higher above the condyloid. Finally, the foramen is not oriented horizontally in anterior-posterior direction, as in the European jaguar, *Panthera gombaszoegensis* (Hemmer 2001), but turns vertically, with a higher opening at the posterior end. The olecranon fossa has an irregular shape and forms a deep cavity in the posterior face. The crista epiconyle lateralis is reminiscent of a lion and a cave lion (van Logchem/Mol 2008), as it is more pronounced in comparison to other *Homotherium* specimens. This feature is interpreted as an indicator for strong muscle attachments, consistent with a large male individual, and not as a defying taxonomic characteristic. Near the proximal edge of the fragment, the onset of the deltaloid ridge is present. The epicondyle and the proximal half of the bone are missing due to damage.

Individual 3 (Jörg, after Jörg Neumann Giesen)

The *Homotherium* neurocranium (**fig. 5**) from layer Schö 13II-4e was found in six larger and several smaller fragments, which were later refitted and fixed. Before restoration, all skull fragments were 3D scanned at Leiden University (the Netherlands), to enable the study of both the external and internal features. The bone preservation is excellent, allowing detailed study of specific features. The endocranial cavity reflects the external shape of the brain, which is of taxonomic and potentially of functional relevance. The third *Homotherium* from Schöningen will therefore potentially contribute to the understanding of the European sabre-toothed cat as a species. At the moment, the research on this specimen is at a preliminary stage; hence, only a short description is presented in this paper.

The neurocranium is triangular in shape, both from a lateral and a posterior view, which is due to the high and pronounced sagittal crest. The occipital area has a rough texture and displays well-developed muscle attachments, related to the strong neck musculature. On the ventral surface, the muscle attachments of the longus capitus muscle are clearly visible in the basisphenoid-occipital suture area. This feature is also described for a *H. latidens* specimen from Incarcal (Antón/Galobart 1999, 777) and the *H. crenatidens* specimen from Senèze (dép. Haute-Loire/F; Ballesio 1963). The mastoid process is very large in comparison to that of modern felids and protrudes laterorostrally (Antón et al. 2004). The external auditory meatus is encapsulated by the frontal border of the mastoid process and the retroarticular process, resulting in a parallel sided slit with a dorsocaudal-ventrorostral direction. The glenoid fossa is positioned ventrally, where in modern felids it is directed ventrorostrally, which is related to the extended gape in sabre-toothed cats. All characteristics described above support the identification of the neurocranium as *Homotherium*.

element	feature	measurements in mm
I1 sin. (ID 26509)	height including root	40.4
	height crown (buccal)	19.6
	maximal width crown	11.3
	length crown (lingual-buccal)	12.3
P4 sin. (ID 26470, ID 26238)	height parastyle	12.2
	anterioposterior length parastyle	8.3
	width parastyle	8.4
	height paracone	19.6
	width paracone	11.5
i3 dex. (ID 26229)	height crown (buccal)	21.8
	maximal width crown	13.6
	length crown (lingual-buccal)	13.8
m1 dex. (ID 26549)	anterioposterior length paraconid	11.5
	height paraconid	14.7
humerus dex. (ID 25885)	smallest breadth diaphyse (SD)	34
	maximal width fossa epicondylar	9
	maximal length fossa epicondylar	27
	minimal thickness epicondylar bridge	5

Tab. 2 Measurements of all *Homotherium* skeletal elements from the site Schö 13II-4, taken with a digital caliper and were rounded to a tenth of a millimeter. Measurements were only performed on the non-fragmentary parts of the skeletal elements.

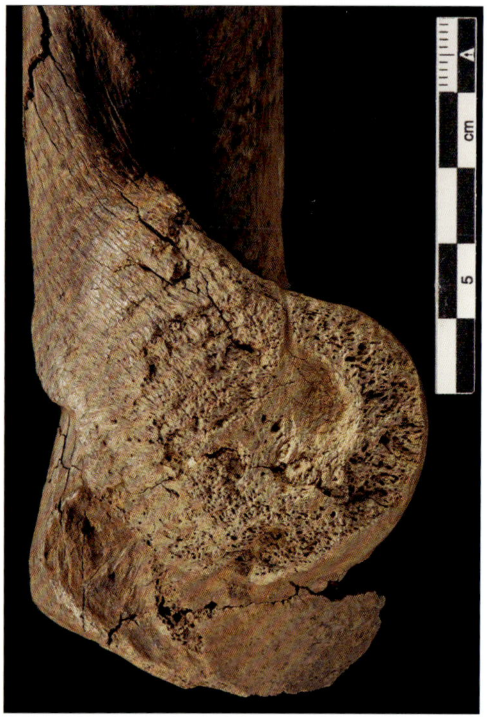

Fig. 6 Schöningen (Lkr. Helmstedt/D): *Homotherium latidens*: medial view of the distal part of a right humerus showing carnivore gnawing marks (ID 25885). – (Photo V. Minkus).

Stratigraphic origin, spatial distribution and taphonomic aspects of the finds

The spatial distribution of the *Homotherium* dental elements from the »Obere Berme« is limited, though slightly dispersed (**fig. 2**), as two refitting parts of the upper P4 were found 70cm apart. Post-depositional processes, either alluvial or trampling, probably caused the distribution of the skeletal and dental elements.

The black, shiny patination of the tooth enamel is similar to other faunal remains at the site, and results from the preservation in humic, organic sediment. The enamel and the roots show hardly any signs of post-mortem weathering; only the root of the upper incisor shows some shallow surface scratches and iron oxide staining. Nonetheless, some of the teeth have been fragmented before final deposition.

Individual 2 (Wolfgang), represented by a distal humerus section, was recovered from the »Speersockel« as an isolated find. It originated from layer Schö 13II-4c, located about 30cm deeper than the base of the main concentration of remains from the Spear Horizon. In this area, a large fissure has been documented throughout the profiles starting at level 13II-4b and continuing to layer Schö 13II-4c. It has repeatedly been documented that due to secondary post-depositional processes, intrusive remains originating from the layer Schö 13II-4b are located in the underlying layer Schö 13II-4c. Hence, we assume that the humerus, too, is intrusive in layer Schö 13II-4c and originates from the overlying archaeological layer. The humerus shows gnawing damage on the distal epiphysis (**fig. 6**) as well as polishing on the proximal edge of the broken diaphysis, probably caused by carnivore licking (**fig. 7, 5**) (Serangeli et al. 2015b). These taphonomic features correspond well with a medium to large-sized carnivore (e.g. wolf or large felid) and indicate the bone still contained nutrients within.

The cranial fragments, belonging to a third individual of *Homotherium*, originated from layer Schö 13II-4e, situated approximately 1 m below the Spear Horizon. The distribution of the cranial fragments is displayed in **figure 8**. The overall bone surface preservation of the skull fragments is good, with microstructures on the muscle attachment areas still present in great detail, indicating minimal weathering. Some fragments, such as the right parietal, show differential preservation with signs of iron oxidation, resulting in a rusty colour. This discoloration could be associated with post-depositional processes within the sediments, resulting in iron oxidation. No obvious taphonomic marks, such as cut marks or gnawing marks have been encountered on the skull fragments, though small unidentifiable scratches are present. Fragmentation appears to have occurred naturally; there is breakage along the thinner parts of the neurocranium, such as the mid regions of the parietal bones, and at the (former) sutures of the basioccipital region. Three specimens show additional cracks that have not resulted in more fragmentation: the left frontal, right parietal and basioccipital. The incomplete cracks and the fragmentation within the small cluster are most probably related to sedimentary pressure. Similar fractures are seen in the complete horse crania from the Spear Horizon site (Schö 13II-4a-4b/c) at Schöningen. The dispersal and earlier separation of the three segments that are not within the cluster must have occurred before deposition. The unweathered bone surface

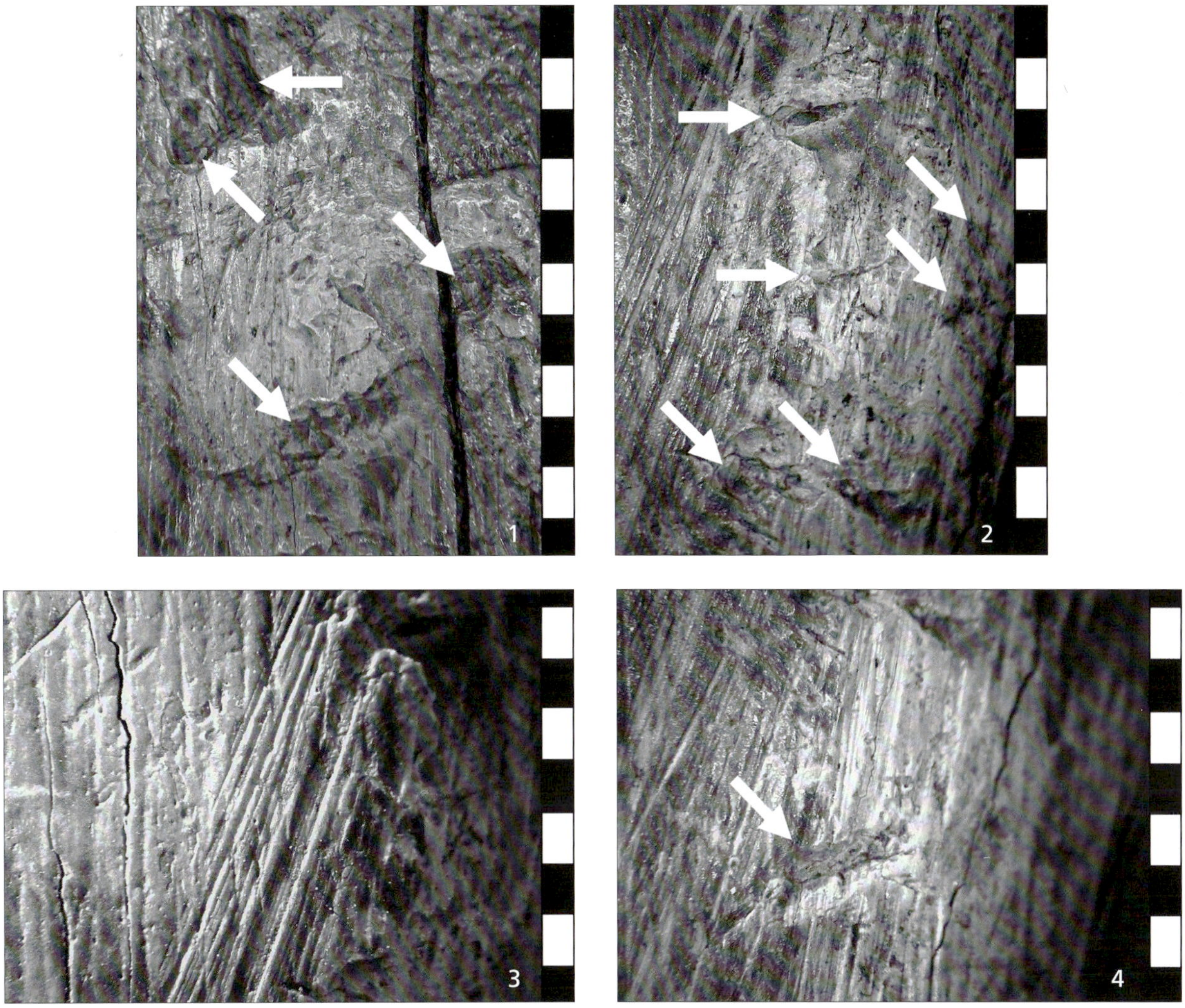

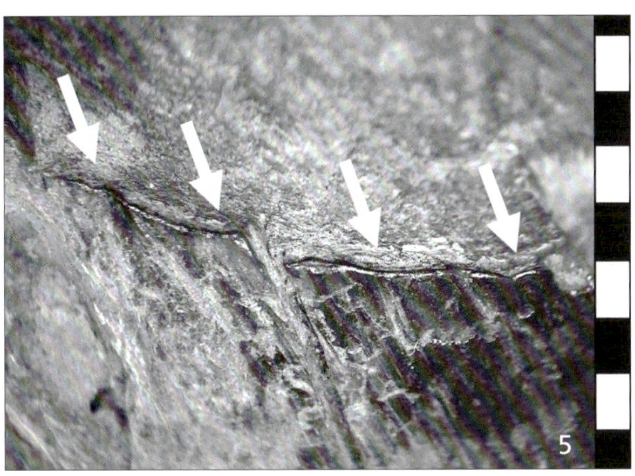

Fig. 7 Schöningen (Lkr. Helmstedt/D): *Homotherium latidens*: medial view of the diaphysis of a right humerus showing percussion marks and the occurrence of underlying, long, shallow scraping marks across the surface of the bone (ID 25885). – **1-4** details of surfaces with scraping marks and traces of repeated impacts (arrows). Scraping marks are mostly parallel, traces of impacts are usually transverse to the long axis of the bone. – **5** details of surfaces with polishing on the proximal edge of the broken diaphysis from gnawing and licking by a wolf-sized canid or large felid. – (Photos overview V. Minkus; details 1-5 J. Serangeli).

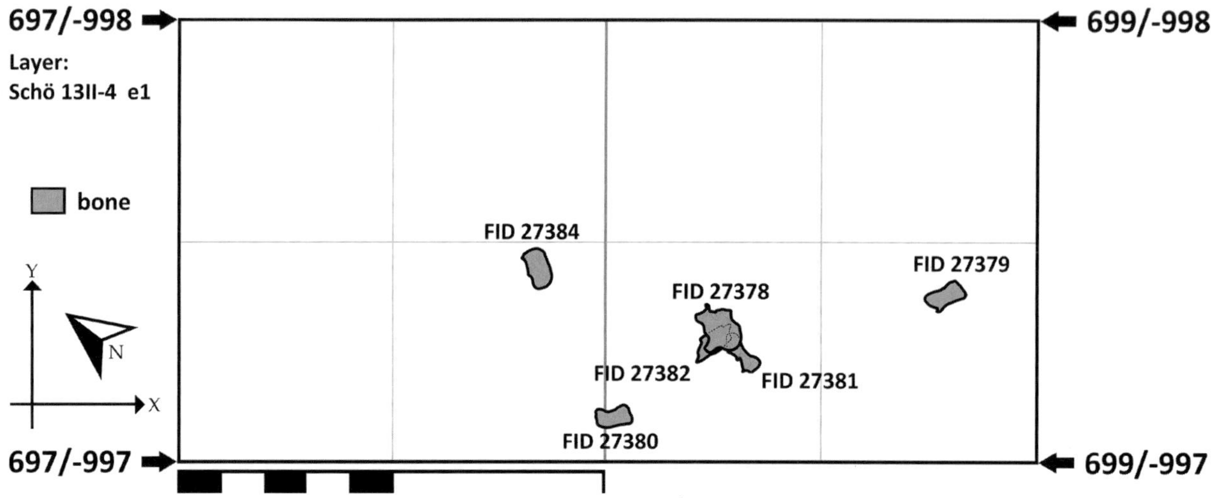

Fig. 8 Schöningen (Lkr. Helmstedt/D): GIS map of the six major fragments of the neurocranium of *Homotherium* (individual 3, Jörg) recovered in the square meters X 697 / Y -997 and X 698 / Y -997. It is important to underline here that the fragments lie in different heights with differences up to 22 cm (ID 27378 = right temporal/parietal bone, Z. 102,07 and ID 27382 = medial part of the right and left parietal and right frontal bone, Z. 101,85). – (Illustration D. Mennella).

preservation without root etching suggests that the cranium was not exposed to the elements over a long period. The distribution is therefore most probably related to trampling or movement within the body of water.

HOMININ MODIFICATION

The humerus shows various taphonomic markers inflicted by both hominin, carnivore and natural actors (**fig. 7**; Serangeli et al. 2015b; van Kolfschoten et al. 2015). The exfoliated surface of some parts of the cortical bone is the result of weathering, prior to other taphonomic damage. On the medial and lateral side, hominin scraping marks and percussion marks from (re)sharpening flint tools are present, all related to its use as a percussor tool (van Kolfschoten et al. 2015; Serangeli et al. 2015b). The hominin scraping marks result from the removal of (remnants of) the periosteum to clean the bone surface, by scraping with the sharp edge of a flint tool in a slightly oblique position over the bone surface with a direction perpendicular to the long axis of the bone. The scraping marks continue straight until the edge of the proximal break of the diaphysis, suggesting this break occurred after the scraping marks were produced. After this cleaning process, the bone was used as a percussor, leaving pits that are fundamentally different from bite scars produced by other animals; the V-shaped section of the marks is indicative of its artificial origin and do not correspond with bite marks, which are usually more shallow and U-shaped. As a final stage of the taphonomic history, the proximal edge of the breakage was polished as a result of carnivores licking the bone. Carnivore gnawing marks are also present on the distal epiphysis, but it cannot be determined if these were inflicted prior to or after hominins utilized the specimen.

The use of bones as a percussor tool at the Schöningen sites is relatively common; van Kolfschoten et al. (2015) describe a multitude of these tools, made of bones from different taxa (e. g., horse, large bovids, red deer and giant deer). The use of (partially) weathered bones for this purpose is not unique, as indicated by a weathered bison radius (ID 7131) that was discovered c. 6 m from the *Homotherium* percussor.

THE YOUNGEST EVIDENCE OF *HOMOTHERIUM LATIDENS* IN EUROPE

The regional extinction of *Homotherium*, and especially the dating of its latest occurrences, is a matter of ongoing debate. Between 1 Ma and 500 000, *Homotherium* first disappeared from the African fossil record, and subsequently around 500 000 in southernmost Europe. Until the end of the 20th century, it was assumed that the presence of the genus *Homotherium* slowly declined in Eurasia until they disappeared around 300 000 years ago. In North America, the genus persisted until the Late Pleistocene, around 12 000 BP as *Homotherium serum* (Turner/Antón 1997; Antón 2013).

Multiple claims have been made for the Late Pleistocene survival of *Homotherium* in Europe. At Kents Cavern (Devon/GB), several dental remains were discovered in association with Upper Palaeolithic artefacts. The teeth were found to be intrusive to the cave sediments, given their dissimilar uranium and fluorine uptake in comparison to other faunal remains from the same location (McFarlane/Lundberg 2013). In 2003, a mandible was dredged from the bottom of the North Sea. It was dated to *c.* 28 000 BP using radiocarbon dating, and is the youngest dated find of *Homotherium* in Europe (Reumer et al. 2003; van Logchem/Mol 2008; Mol et al. 2008). The exact stratigraphic origin of the find is unknown, although the preservation and mineralisation correspond with other Late Pleistocene faunal remains from the area. It was originally ascribed to *Homotherium latidens*, other authors however highlight its similarities in morphology to the Late Pleistocene species *H. serum* (Antón et al. 2014).

Excavations in the municipality of Mealhada (distr. Aveiro/P) during the 19th century yielded an astragalus, ascribed to *Homotherium latidens* (Antunes 1986). The river terrace sediments where the faunal remains were recovered were initially correlated with the Riss-Würm (= Eemian, MIS 5e) interglacial (Antunes 1986), but these were later correlated to an interstadial of the Riss (= Saalian; Antunes/Cardoso/Faure 1988; Cunha-Ribeiro 1999). Dental remains discovered in layer Va of Artenac Cave (dép. Charente/F) consist of a canine, an upper P3 and P4. The chronological assignment of these finds is problematic, as only the adjacent layers have been dated using the uranium/thorium dating technique: the overlying stalagmite floor (PL V) is dated to *c.* 110 000/100 000 years ago, whereas the underlying layer (PL-IV) is 345 000/178 000 years old (Delanges et al. 1999). The associated faunal assemblage shows similarities with other fauna complexes attributed to the Mindel-Riss interglacial or the beginning of the Riss (Beden et al. 1984) and can therefore either be correlated with marine oxygen isotope stage (MIS) 7, MIS 9 or MIS 11.

At Steinheim an der Murr (Lkr. Ludwigsburg/D), a canine of *Homotherium* was found in a layer dated to the Holsteinian Interglacial or the beginning of the Saalien (Adam 1961). In the very same layer, the Steinheim skull, hominin cranial remains most likely belonging to *Homo heidelbergensis* were recovered, indicating the contemporaneity of both taxa.

The finds from Schöningen originate from a layer dated to MIS 9 (320 000-300 000 BP; Sierralta/Frechen/Urban 2012; Richter/Krbetscheck 2015). Their well-documented archaeological context adds to their scientific value, as one of the latest occurrences of *Homotherium* within Europe.

DISCUSSION AND CONSEQUENCES FOR HUMAN EVOLUTION

Apart from Schöningen, there are other archaeological sites in Europe where remains of sabre-toothed cats (*Homotherium* or *Megantereon*) were found: Czech Republic: Stránská Skála (okr. Brno-město/CZ), *c.* 780 000-400 000 (Sardella/Iurino 2012); Spain: la Trinchera Dolina de Atapuerca (prov. Burgos/E), layer TD5, 900 000-800 000 (Garcia/Arsuaga 1999) and layer TD10, *c.* 400 000 (Antón/Galobart/Turner 2005);

Italy: Pirro Nord (prov. Foggia/I), 1700000-1300000 (Gliozzi et al. 1997; Arzarello et al. 2007) and Fontana Ranuccio (prov. Frosinone/I), *c.* 450000 (Bittiddu et al. 1979; Gliozzi et al. 1997); Germany: Steinheim an der Murr, *c.* 400000-300000 (Adam 1961) and Mauer bei Heidelberg, 600000-500000 (Freudenberg 1929); Hungary: Vértesszőlős (Kom. Komárom-Esztergom/H), *c.* 600000-300000 (Sardella/Iurino 2012); Greece: Petralona (Chalkidiki/GR), *c.* 600000-400000 (Kurtén/Poulianos 1980). Nonetheless, the co-occurrence of both hominin fossils or artefacts, and saber-toothed cat remains at a site, or even within the same archaeological horizon, does not necessarily prove the contemporaneity of the two taxa; at Kent's Cavern in England, the *Homotherium latidens* dental remains were found to be intrusive to the archaeological sediments, and are now interpreted as potential Upper Palaeolithic trade goods (McFarlane/Lundberg 2013).

By reconstructing the diet of extinct carnivore species with no modern analogue like *Homotherium*, we can hypothesize about the degree of overlap in prey-preference with hominins and the potential for competition over resources. The only direct evidence for the prey-preference of *Homotherium* originates from Friesenhahn Cave (Texas/USA). Here, an accumulation of young proboscidean remains with gnawing marks and the remains of several individuals of sabre-toothed cats with a wide age range were found (Marean/Ehrhardt 1995). If indeed *Homotherium* was actively hunting or scavenging from these large-sized prey animals, most of the large mammal taxa found in Schöningen would also fall into this range; giant deer, bovines, rhinos, large Pleistocene equids and proboscideans have all been discovered at Schöningen, and the majority of them contain hominin butchering marks as well. Therefore, the potential overlap in prey taxa between hominins and *Homotherium* should not be underestimated. In the past, hominins were mainly ascribed a passive role in this relationship, as practicing kleptoparasitism, eating the »leftovers« of *Homotherium* kills (Marean/Ehrhardt 1995; Antón/Galobart/Turner 2005; Antón 2013). The super-positioning of cutmarks and gnawing marks is used as an indicator of primary access to a carcass. In the case of Schöningen, the predominance of gnawing marks overlaying cut marks argues for primary access by hominins (Voormolen 2008; Starkovich/Conard 2015). This pattern is also described for other Lower Palaeolithic, such as Boxgrove in England (West Sussex/GB; Smith 2012; 2013).

The occurrence of *Homotherium* in the archaeological site Schö 13II-4 requires a re-interpretation of the wooden spears solely as hunting weapons; self-defence against large carnivores might have also played a role for carrying weapons while acquiring food through hunting or scavenging.

CONCLUSION

The Schöningen remains are the best-documented finds of *Homotherium* within an archaeological setting. They originate from deposits dated to MIS 9 and are therefore one of the youngest findings of the species *H. latidens* in Europe. Other finds of similar or assumed younger age often originate from poorly documented excavations (e.g. Adam 1961; Antunes 1986) or show more similarities with the American species *H. serum* (e.g. Reumer et al. 2003).

The *Homotherium* humerus with flint percussion marks indicates that the European saber-toothed cat and hominins shared the Middle Pleistocene landscape at around 300000 years ago. It highlights the potential competition between the two species, as they probably hunted a similar suite of prey taxa. And it sheds new light on the technological adaptations (e.g. the Schöningen spears) hominins were using, not only as active weapons for hunting, but possilbly also for self-defence.

ACKNOWLEDGEMENTS

The discovery and the scrupulous documentation of all the remains presented in this paper are, at first, the result of the daily work and the years of experience of all members of the Schöningen excavation team: Wolfgang Mertens, Wolfgang Berkemer, Neil Haycock, Martin Kursch, Jörg Neumann Giesen and Dennis Mennella (Eberhard Karls Universität Tübingen) as well of the technician Jens Lehmann (Niedersächsisches Landesamt für Denkmalpflege, Hannover). – Many thanks to Nicholas J. Conard (Eberhard Karls Universität Tübingen), Thomas Terberger (Niedersächsisches Landesamt für Denkmalpflege, Hannover) and Felix Hillgruber (Paläon) for their support and for the coordination of the sabre-toothed cat-workshop in Schöningen in June 2015. – Many thanks for the help with the identification of the first *Homotherium* tooth to Alain Argant (Université d'Aix-Marseille, Laboratoire méditerranéen de préhistoire Europe Afrique, UMR 7269), Juliá Maroto (Universidad de Girona), Marina Sotnikova (Russian Academy of Sciences, Moscow), Susanne Münzel (Eberhard Karls Universität Tübingen), and to Neil Haycock (Eberhard Karls Universität Tübingen) for his linguistic support.

REFERENCES

Adam 1961: K. D. Adam, Die Bedeutung der pleistozänen Säugetier-Faunen Mitteleuropas für die Geschichte des Eiszeitalters. Stuttgarter Beiträge zur Naturkunde 78, 1961, 1-34.

Antón 2013: M. Antón, Sabertooth (Bloomington, Indianapolis 2013).

Antón/Galobart 1999: M. Antón / A. Galobart, Neck Function and Predatory Behaviour in the Scimitar Toothed Cat *Homotherium latidens* (Owen). Journal of Vertebrate Palaeontology 19/4, 1999, 771-784.

Antón/Galobart/Turner 2005: M. Antón / A. Galobart / A. Turner, Co-existence of Scimitar-toothed Cats, Lions and Hominins in the European Pleistocene. Implications of the Post-cranial Anatomy of *Homotherium latidens* (Owen) for Comparative Palaeoecology. Quaternary Science Reviews 24, 2005, 1287-1301.

Antón et al. 2004: M. Antón / M. J. Salesa / J. F. Pastor / I. M. Sánchez / S. Fraile / J. Morales, Implications of the Mastoid Anatomy of Larger Extant Felids for the Evolution and Predatory Behaviour of Sabre-toothed Cats (Mammalia, Carnivora, Felidae). Zoological Journal of the Linnean Society 140, 2004, 207-221.

2014: M. Antón / M. J. Salesa / A. Galobart / Z. J. Tseng, The Plio-Pleistocene Scimitar-toothed Felid Genus *Homotherium* Fabrini, 1890 (Machairodontinae, Homotherini): Diversity, Palaeogeography and Taxonomic Implications. Quaternary Science Reviews 96, 2014, 259-268.

Antunes 1986: M. T. Antunes, Acerca de um osso do plistocénico da Mealhada: presença de um »tigre dente de sabre«, *homotherium latidens* (OWEN, 1846). Ciências da Terra 8, 1986, 43-54.

Antunes/Cardoso/Faure 1988: M. T. Antunes / J. L. Cardoso / M. Faure, Présence de *Hippopotamus incognitus* au Portugal et remarques sur les sites quaternaires de Mealhada. Comunicações dos Serviços Geológicos de Portugal 74, 1988, 165-172.

Arzarello et al. 2007: M. Arzarello / F. Marcolini / G. Pavia / M. Pavia / C. Petronio / M. Petrucci / L. Rook / R. Sardella, Evidence of earliest human occurrence in Europe: the site of Pirro Nord (Southern Italy). Naturwissenschaften 94, 2007, 107-112.

van Asperen 2004: E. N. van Asperen, The large mammal fauna of Schöningen 13II-4 [MA thesis, Univ. Leiden 2004].

Ballesio 1963: R. Ballesio, Monographie d'un Machairodus du gisement villafranchien de Senéze: *Homotherium crenatidens* fabrini. Travaux du Laboratoire de Géologie de la Faculté de Sciences de Lyon N.S. 9 (Saint-Étienne 1963).

Beden et al. 1984: M. Beden / L. De Bonis / M. Brunet / J. F. Tournepiche, Première découverte d'un félin machairodonte dans le Pleistocene moyen des Charentes. Comptes-Rendus de l'Académie des Sciences de Paris 289, 1984, 241-244.

Bittiddu et al. 1979: I. Bittiddu / P. F. Cassoli / F. Radicati di Brozolo / A. G. Segre / E. Segre Naldini / I. Villa, Anagni, a K-Ar dated Lower and Middle Pleistocene Site, Central Italy: Preliminary Report. Quaternaria Nova 21, 1979, 53-71.

Cunha-Ribeiro 1999: J. P. Cunha-Ribeiro, A estação paleolítica da Mealhada nos 120 anos de estudo do Acheulense em Portugal. O Arqueólogo Português Sér. IV, 13/15, 1995-1997 (1999), 35-52.

Delagnes et al. 1999: A. Delagnes / J.-F. Tournepiche / D. Armand / E. Desclaux / M.-F. Diot / C. Ferrier / V. Le Fillâtre / B. Vandermeersch / J.-J. Bahain / J.-G. Bordes / C. Falguères / L. Ayliffe / L. Froget / N. Mercier / H. Valladas, Le gisement Pléistocène moyen et supérieur d'Artenac (Saint-Mary, Charente): premier bilan interdisciplinaire. Bulletin de la Société Préhistorique Française 96/4, 1999, 469-496.

Ficcarelli 1979: G. Ficcarelli, The Villafranchian machairodonts of Tuscany. Palaeontographia Italica 71 (N.S. 41), 1979, 17-26.

Freudenberg 1929: W. Freudenberg, Ein Säbeltiger und ein großer Anthropoide(?) aus den Sanden von Mauer a.d. Elsenz. Zeitschrift der Deutschen Geologischen Gesellschaft 81, 1929, 291.

Galobart et al. 2003: A. Galobart / J. Pons-Moyà / M. Antón / J. Maroto, Descripción del material de *Homotherium latidens* (Owen) de los yacimientos del Pleistoceno inferior de Incarcal (Girona, NE de la Península Ibérica). Paleontologia i Evolució 34, 2003, 99-141.

Garcia/Arsuaga 1999: N. Garcia / J. L. Arsuaga, Carnivores from the Early Pleistocene hominid-bearing Trinchera Dolina 6 (Sierra de Atapuerca, Spain). Journal of Human Evolution 37, 1999, 415-430.

Gliozzi et al. 1997: E. Gliozzi / L. Abbazzi / A. Azzaroli / L. Caloi / L. Capasso Barbato / G. Di Stefano / D. Esu / G. Ficcarelli / O. Girotti / T. Kotsakis / F. Masini / P. Mazza / C. Mezzabotta / M. R. Palombo / C. Petronio / L. Rook / B. Sala / R. Sardella / E. Zanalda / D. Torre, Biochronology of Selected Mammals, Molluscs and Ostracods from the Middle Pliocene to the Late Pleistocene in Italy. The State of the Art. Rivista Italiana di Paleontologia e Stratigrafia 90, 1997, 369-388.

Heinrich/van Kolfschoten 2007: W. D. Heinrich / T. van Kolfschoten, Erster Skelettfund von *Trogontherium cuvieri* (Altbiber). Eine neu entdeckte Warmzeit in Schöningen: Das Reinsdorf-Interglazial. In: H. Thieme (ed.), Die Schöninger Speere – Mensch und Jagd vor 400.000 Jahren (Stuttgart 2007) 118-123.

Hemmer 2001: H. Hemmer, Die Feliden aus dem Epivillafranchium von Untermaßfeld. In: R.-D. Kahlke (ed.), Das Pleistozän von Untermassfeld bei Meiningen (Thüringen), Teil 3. Monographien des RGZM 40, 3 (Mainz 2001) 699-782.

van Kolfschoten 2014: T. van Kolfschoten, The Palaeolithic Locality Schöningen (Germany): A Review of the Mammalian Record. Quaternary International 326-327, 2014, 469-480.

van Kolfschoten et al. 2015: T. van Kolfschoten / S. Parfitt / J. Serangeli / S. Bello, Lower Palaeolithic Bone Tools from the Schöningen Spear Horizon (Germany). Journal of Human Evolution 89, 2015, 226-263.

Kurtén/Poulianos 1980: B. Kurtén / A. N. Poulianos, Fossil Carnivora of Petralona Cave: Status of 1980. Anthropos 8, 1980, 9-56.

van Logchem/Mol 2008: W. van Logchem / D. Mol, De vroeg-pleistocene sabeltandkat, *Homotherium crenatidens* (WEITHOFER, 1889), voor het eerst opgevist van de bodem van de Noordzee. Cranium 25/2, 2008, 3-16.

Marean/Ehrhardt 1995: C. W. Marean / C. L. Ehrhardt, Paleoanthropological and Paleoecological Implications of the Taphonomy of a Sabertooth's Den. Journal of Human Evolution 29, 1995, 515-547.

McFarlane/Lundberg 2013: D. A. McFarlane / J. Lundberg, On the Occurrence of the Scimitar-toothed Cat, *Homotherium latidens* (Carnivora; Felidae), at Kents Cavern, England. Journal of Archaeological Science 40, 2013, 1629-1635.

Mol et al. 2008: D. Mol / W. van Logchem / K. van Hooijdonk / R. Bakker, The saber-toothed cat of the North Sea (Norg 2008).

Naples/Martin/Babiarz 2011: V. L. Naples / L. D. Martin / J. P. Babiarz, The Other Saber-tooths. Scimitar-tooth Cats of the Western Hemisphere (Baltimore 2011).

Reumer et al. 2003: J. W. F. Reumer / L. Rook / K. van der Borg / K. Post / D. Mol / J. de Vos, Late Pleistocene Survival of the Saber-toothed Cat *Homotherium* in Northwestern Europe. Journal of Vertebrate Paleontology 23/1, 2003, 260-262.

Richter/Krbetschek 2015: D. Richter / M. Krbetschek, The Age of the Lower Palaeolithic Occupation at Schöningen. Journal of Human Evolution 89, 2015, 46-56.

Rincón/Prevosti/Parra 2011: A. D. Rincón / F. J. Prevosti / G. E. Parra, New Saber-toothed Cat Records (Felidae: Machairodontinae) for the Pleistocene of Venezuela, and the Great American Biotic Interchange. Journal of Vertebrate Paleontology 31/2, 2011, 468-478.

Sardella/Iurino 2012: R. Sardella / D. A. Iurino, The Latest Early Pleistocene Sabertoothed Cat *Homotherium* (Felidae, Mammalia) from Monte Peglia (Umbria, central Italy). Bollettino della Società Paleontologica Italiana 51/1, 2012, 15-22.

Serangeli et al. 2015a: J. Serangeli / U. Böhner / T. van Kolfschoten / N. J. Conard, Overview and new results from large-scale excavations in Schöningen. Journal of Human Evolution 89, 2015, 27-45.

2015b: J. Serangeli / T. van Kolfschoten / B. Starkovitch / I. Verheijen, The European Saber-toothed cat (*Homotherium latidens*) found in the »Spear Horizon« at Schöningen (Germany). Journal of Human Evolution 89, 2015, 172-180.

Sierralta/Frechen/Urban 2012: M. Sierralte / M. Frechen / B. Urban, ^{230}Th/U dating results from opencast mine Schöningen. In: K.-E. Behre (ed.), Die chronologische Einordnung der paläolithischen Fundstellen von Schöningen. The chronological setting of the Palaeolithic sites of Schöningen. Forschungen zur Urgeschichte im Tagebau von Schöningen 1 (Mainz 2012) 143-154.

Smith 2012: G. M. Smith, Hominin-Carnivore Interaction at the Lower Palaeolithic Site of Boxgrove, UK. Journal of Taphonomy 10, 2012, 373-394.

2013: G. M. Smith, Taphonomic Resolution and Hominin Subsistence Behaviour in the Lower Palaeolithic: Differing Data Scales and Interpretive Frameworks at Boxgrove and Swanscombe (UK). Journal of Archaeological Science 40, 2013, 3754-3767.

Sotnikova/Titov 2009: M. Sotnikova / V. Titov, Carnivora of the Tamanian faunal unit (the Azov Sea area). Quaternary International 201, 2009, 43-52.

Starkovich/Conard 2015: B. Strakovich / N. J. Conard, Bone Taphonomy of the Schöningen »Spear Horizon South« and its Implications for Site Formation and Hominin Meat Provisioning. Journal of Human Evolution 89, 2015, 154-171.

Tournepiche 1996: J. F. Tournepiche, Les grands mammifères pléistocènes de Poitou-Charente. Paléo 8, 1996, 109-141.

Turner/Antón 1997: A. Turner / M. Antón, The Big Cats and their Fossil Relatives (New York 1997).

Voormolen 2008: B. Voormolen, Ancient Hunters, Modern Butchers. Schöningen 13II-4, a Kill-Butchery Site Dating from the Northwest European Lower Paleolithic. Journal of Taphonomy 6/2, 2008, 71-247.

SUMMARY / ZUSAMMENFASSUNG

The Remains of the European Sabre-toothed Cat (*Homotherium latidens*) Found in the Spear Horizon at Schöningen (Germany)

The discovery of remains of a European sabre-toothed cat (*Homotherium latidens*) in the Spear Horizon is one of the most important discoveries of the last years at Schöningen (Lkr. Helmstedt/D). Here we present the remains of three individuals: (1) seven teeth and associated postcranial elements, (2) a distal humerus fragment and (3) a neurocranium. The remains of the first two individuals were found in the archaeological site Schöningen 13II-4, layers 4a-4bc, the so-called Spear Horizon, the approximately 300 000-320 000-year-old layers where the world's oldest preserved wooden spears were discovered. The teeth were found in close proximity to each other and all belong to a c. 2-3-year-old sub-adult individual. Conversely, the humerus is from a large and robust individual, probably a male. It displays evidence of hominin scrape and impact marks that indicate it was used as a percussor for working flint. To date, this is the only known bone of a *Homotherium* that has been modified and used by hominins. The cranial fragments of the third individual were excavated in 2015 from the Schöningen 13II-4, layer 4e, situated approximately 1 m below the other finds. Six larger fragments were 3D documented and refitted, and together they form the neurocranium of an adult specimen.

The *Homotherium* remains from Schöningen have an age of c. 300 000-320 000 years and are therefore amongst the youngest specimens of *Homotherium latidens* from Europe. The presence of this large carnivore competitor would certainly have had a major impact on the existence of late Middle Pleistocene hominins. From this perspective, one can hypothesize that the wooden spears were not only used for hunting large mammals, but probably also as a weapon for self-defence.

Die Überreste des Europäischen Säbelzahntigers (*Homotherium latidens*) im Speerhorizont bei Schöningen (Deutschland)

Der Fund von Überresten einer Europäischen Säbelzahnkatze (*Homotherium latidens*) im Speerhorizont ist eine der wichtigsten Entdeckungen der letzten Jahre bei Schöningen (Lkr. Helmstedt/D). Hier präsentieren wir die Überreste von drei Individuen: (1) sieben Zähne und zugehörige postcraniale Elemente, (2) ein distales Humerusfragment und (3) ein Neurocranium. Die Überreste der ersten beiden Individuen wurden in der archäologischen Stätte Schöningen 13II-4, Schichten 4a-4bc, dem sogenannten Speerhorizont, gefunden – den etwa 300 000-320 000 Jahre alten Schichten, in denen die ältesten erhaltenen Holzspeere der Welt entdeckt wurden. Die Zähne wurden in unmittelbarer Nähe zueinander gefunden und gehören alle zu einem ca. 2-3 Jahre alten subadulten Individuum. Der Oberarmknochen hingegen stammt von einem großen und kräftigen Individuum, wahrscheinlich einem männlichen Tier. Er weist Schürf- und Schlagspuren von Homininen auf, die darauf hindeuten, dass er als Schlaginstrument zur Bearbeitung von Feuerstein verwendet wurde. Bislang ist dies der einzige bekannte Knochen eines *Homotherium*, der von Homininen verändert und benutzt wurde. Die Schädelfragmente des dritten Individuums wurden 2015 in Schöningen 13II-4, Schicht 4e, etwa 1 m unter den anderen Funden ausgegraben. Sechs größere Fragmente wurden 3D-dokumentiert und nachbearbeitet und bilden zusammen das Neurocranium eines erwachsenen Exemplars.

Die *Homotherium*-Reste aus Schöningen haben ein Alter von ca. 300 000-320 000 Jahren und gehören damit zu den jüngsten Exemplaren von *Homotherium latidens* aus Europa. Die Anwesenheit dieses großen fleischfressenden Konkurrenten hätte sicherlich einen großen Einfluss auf die Existenz der spätmittelpleistozänen Homininen gehabt. Unter diesem Gesichtspunkt kann man annehmen, dass die Holzspeere nicht nur für die Jagd auf große Säugetiere, sondern wahrscheinlich auch als Waffe zur Selbstverteidigung verwendet wurden.

MAURICIO ANTÓN

BEHAVIOUR OF *HOMOTHERIUM* IN THE LIGHT OF MODERN AFRICAN BIG CATS

How can we get a picture of the sabretooth cat *Homotherium* as a living animal? We use a methodology of paleobiological reconstruction that combines the study of fossil remains with that of their modern counterparts in order to propose hypotheses about the unpreserved anatomy and behaviour of the extinct species. In this contribution I explain how the study of the functional anatomy of the fossil species is complemented by comparative anatomy, the dissection of extant carnivores, and the direct observation of living big cats in the African savannah in order to achieve a balanced view of the hypothetical appearance, action and ecology of *Homotherium*.

ANATOMICAL RECONSTRUCTION

The first step of the process of paleobiological reconstruction is the restoration of the physical appearance of *Homotherium*. Any image that shows extinct animals in action must be the result of a careful process, based on a solid methodology. The method that we use today was originally devised by Georges Cuvier and decades later it was independently recreated by Charles Knight. More recent developments have refined the phylogenetic aspects of reconstruction (Antón/Sánchez 2004; Bryant/Russell 1992; Witmer 1995). At its core, the method consists in a step-by-step restoration of tissue from the inside out, as in a reverse dissection. First we restore the missing parts of the skeleton and articulate it in a life-like pose. Then we add succesive layers of soft tissue, starting with deep muscles and using marks in the attachment areas of the bones as a guide, and ending with speculation about fur patterns (**fig. 1**).
This procedure is only possible thanks to the existence of relatively complete fossil samples. As early as 1914, Freudenberg described parts of the skeleton of *Homotherium* on the basis of the sample from Hundsheim Cave (Bez. Bruck an der Leitha/A) in Austria, and in 1925 Schaub referred to that sample in order to remark the contrasting skeletal morphologies of *Megantereon* and *Homotherium,* the two best-known sabretooth genera from Plio-Pleistocene Europe (Freudenberg 1914; Schaub 1925). But only with the description of the nearly complete skeleton from Senèze (dép. Haute-Loire/F) did a more complete picture of *Homotherium* emerge. In 1963 Ballesio described that skeleton and showed that *Homotherium* differed greatly from the better known, classic sabretooth *Smilodon*. The latter was a short-legged, robust and heavy animal, while *Homotherium* was revealed as comparatively light and long-legged (Ballesio 1963).
Our own research efforts have brought together the information from Senéze and Hundsheim with the rich sample from the Spanish site of Incarcal (prov. Girona/E) to create an updated reconstruction of *Homotherium*. The animal's neck was long, strong and flexible, while the back was short and the tail was not much more than a stump (Antón/Galobart 1999; Antón et al. 2005). We showed that *Homotherium* had narrow forelimbs intermediate between those of lions and cheetahs, with relatively small, semi-retractable claws and a huge dewclaw (**fig. 2**).

In: Nicholas J. Conard · Henning Hassmann · Kurt Felix Hillgruber · Jordi Serangeli · Thomas Terberger (eds), The *Homotherium* Finds from Schöningen 13II-4. Man and Big Cats of the Ice Age. Forschungen zur Urgeschichte aus dem Tagebau von Schöningen 4 (Mainz 2022). DOI: https://doi.org/10.11588/propylaeum.1006.c13519

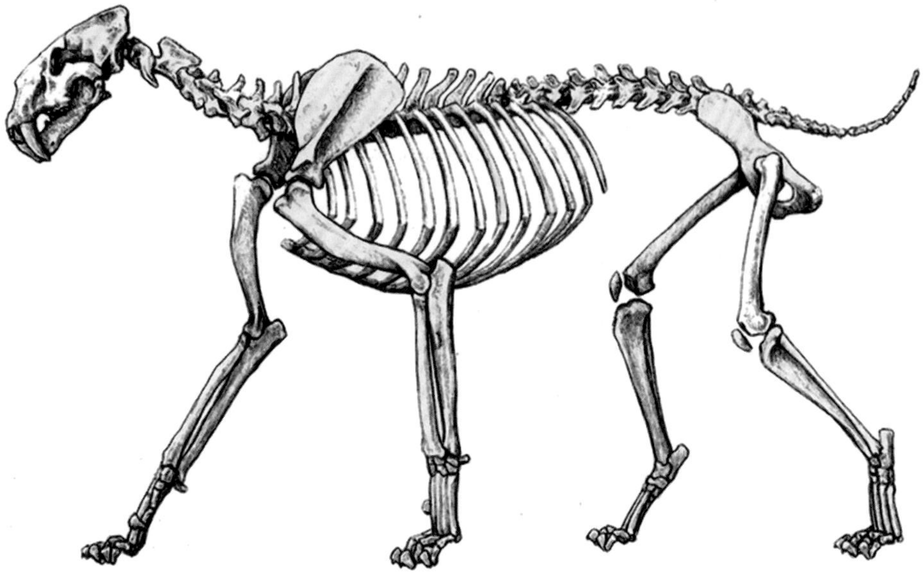

1

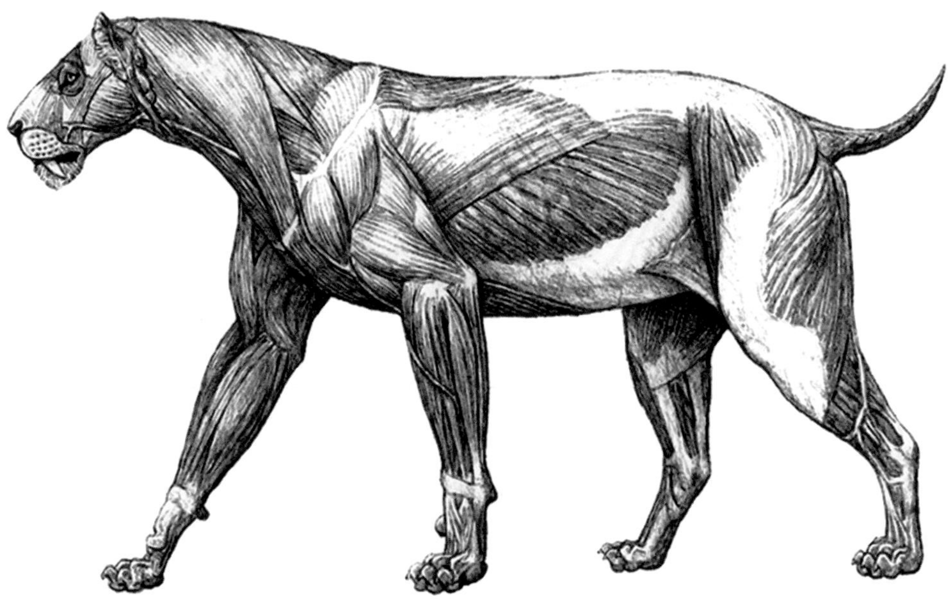

2

3

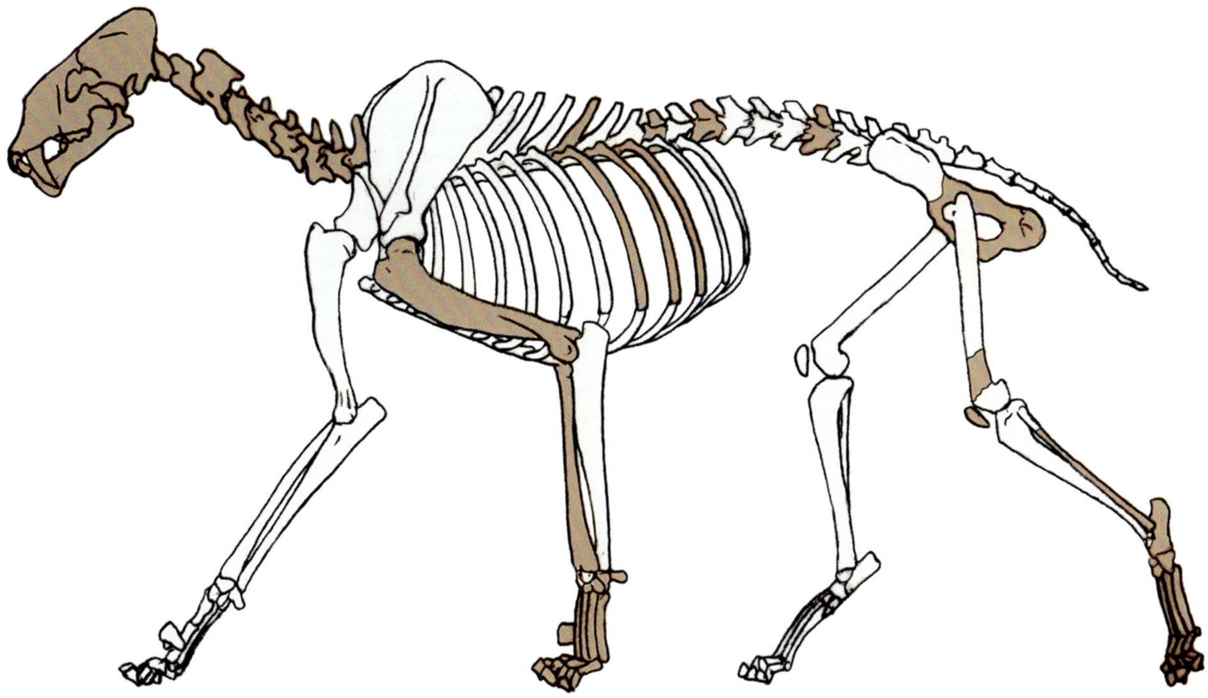

Fig. 2 Skeleton of *Homotherium* with the sections preserved at the fossil site of Incarcal shown in a darker colour. – (Artwork M. Antón).

Combining these data with those about the other sabretooth cats from the Pliocene and Pleistocene of the Old World we could make an overall comparison of sizes and proportions (**fig. 3**). The differences in body size and proportions between species in turn imply a partitioning of available resources. But in order to know how this partition would take place, it is essential to interpret the morphology of sabretooths in functional terms, and to look again at the morphology and ecology of modern big cats for reference (Turner/Antón 1999).

BIOMECHANICAL IMPLICATIONS

The image that emerges from our physical reconstruction of *Homotherium* shows a mosaic of similarities and differences with other sabretooths but also with modern big cats. What are the implications of that mosaic in terms of the locomotor and predatory behaviour of *Homotherium*? In order to know that, we needed to interpret the anatomy of *Homotherium* in the light of the relationships between morphology and locomotion in modern cats. To explore those relationships we conducted dissections and we also analysed video footage to study the gaits of living cats.

The detailed comparison of individual bones between *Homotherium* and lions shows differences that influence locomotion (Antón et al. 2005). The short, less asymmetrical second phalanges and the small claws point to more »dog-like« feet. Less retractable claws provide better traction, but they get worn and thus

Fig. 1 Three steps of the reconstruction of *Homotherium latidens*: **1** reconstruction of the skeleton in a life-like pose. – **2** reconstruction of the musculature. – **3** reconstructed life appearance, with inferred fur pattern. – (Artwork M. Antón).

1

2

3

are less efficient for grasping prey. The narrow distal humerus with a more vertical olecranon fossa implies a more parasaggital trajectory of the forearm and wrist during the walk, again resembling the efficient walk of dogs more than that of cats with its marked supination of the forepaw as the limb moves forward. Analysis and comparison of my footage of free-ranging lions and hyaenas in Africa provides clear evidence of the walking styles associated with different forelimb morphologies (**fig. 4**: frames from videos of carnivores walking in front view). It is likely that the walk of *Homotherium* would be intermediate between those extremes. One interesting observation is the fact that the locomotion pattern of primitive sabretooths, such as *Machairodus* from the late Miocene, would be similar to that of extant cats, while the pattern of *Homotherium* is more derived and subtly different (Antón 2013).
Concerning gait, some old artistic reconstructions showed *Homotherium* as a plantigrade animal, on the basis of some features of the hind limb that could fit with a plantigrade stance (**fig. 5**). But Ballesio's study clearly proved the contrary, and our own observations add further evidence to confirm a fully digitigrade hind limb. In fact, apparent »plantigrade« adaptations in *Homotherium* would not be related to its walking gait, but to its posture while bringing down and immobilising struggling prey (Antón et al. 2005). Body proportions and morphology indicate that the animal could run at moderate speeds more efficiently than smilodontines, and even lions, and for greater distances (**fig. 6**).
Our dissections of modern cats further clarified aspects of the sabretooth's killing bite. We identified the muscles attaching to precise areas of the skull and cervical vertebrae in modern cats, allowing a more accurate inference of the musculo-skeletal system in sabretooth felids (Antón et al. 2004). The upper canines of sabretooths, and especially those of *Homotherium*, were longer and more flattened, and thus more fragile than those of modern cats. This requires a greater gape, if only to achieve a similar clearance between canine tips (and thus for biting at prey of similar size), and also a greater contribution of neck muscles to the bite, since they can bring the whole head down providing extra strength for the penetration of the canines into the flesh of prey (**fig. 7**). But it also becomes necessary to avoid hitting bone during the bite in order to prevent canine breakage, a requisite that would lead the sabretooths to avoid killing smaller prey, as hitting bone is more likely with smaller animals. Also, sabretooths need to control the struggles of the prey in order to avoid sudden motions that could break the canines. In some sabretooth species this is achieved by developing very robust and powerful bodies with enormously strong grasping forepaws. But that was not the case with *Homotherium*, where a somewhat different solution was apparently achieved.

PREY SIZE

It seems clear that *Homotherium*'s locomotion shared some aspects of cats and hyaenas, and that its craniodental adaptations point to the quick killing of large prey. But the resulting picture is somewhat contradictory because precisely the adaptations for sustained running that it shares with hyaenas pose limitations for bringing down the large prey that its craniodental anatomy was adapted to kill. Observations of modern carnivores give further indications about *Homotherium*'s habits and help to clarify those apparent contradictions.

Fig. 3 Comparison of body proportions and size in three species of machairodontine felids from the Pliocene and Pleistocene of Europe: **1** *Dinofelis diastemata*. – **2** *Megantereon cultridens*. – **3** *Homotherium latidens*. – All drawn to the same scale. Shoulder height of *Homotherium* about 1.1 m. – (Artwork M. Antón).

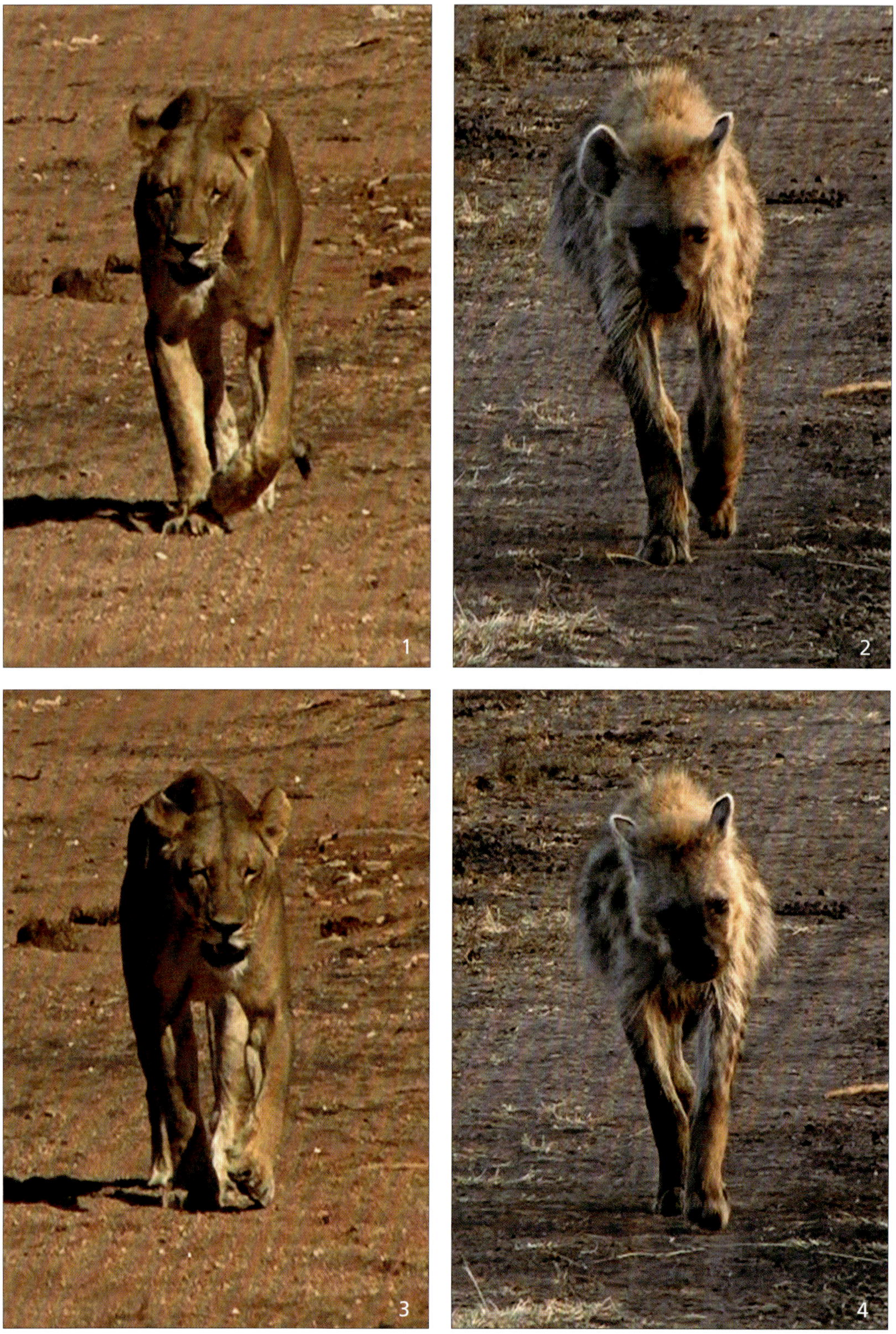

Fig. 4 Frames from videos of walking lions and spotted hyaenas in front view, showing differences in locomotion related to forelimb osteology: **1** lioness with left forelimb advanced and wrist flexed; notice that the forepaw is facing inwards and supinating. – **2** hyaena in the same stage of the walk; notice that the forearm and paw are aligned in the same vertical axis, with no appreciable supination of the paw. – **3** lioness with left forepaw just contacting the ground; notice that the forepaw is supinated so that the palm faces inward and the fifth digit touches the ground first. – **4** hyaena in the same stage of the walk; notice that the paw is not supinated so that the four digits contact the ground simultaneously. The intermediate forelimb morphology of *Homotherium* suggests that its locomotion would be somewhere in between the extremes shown in this figure. – (Filmed by M. Antón).

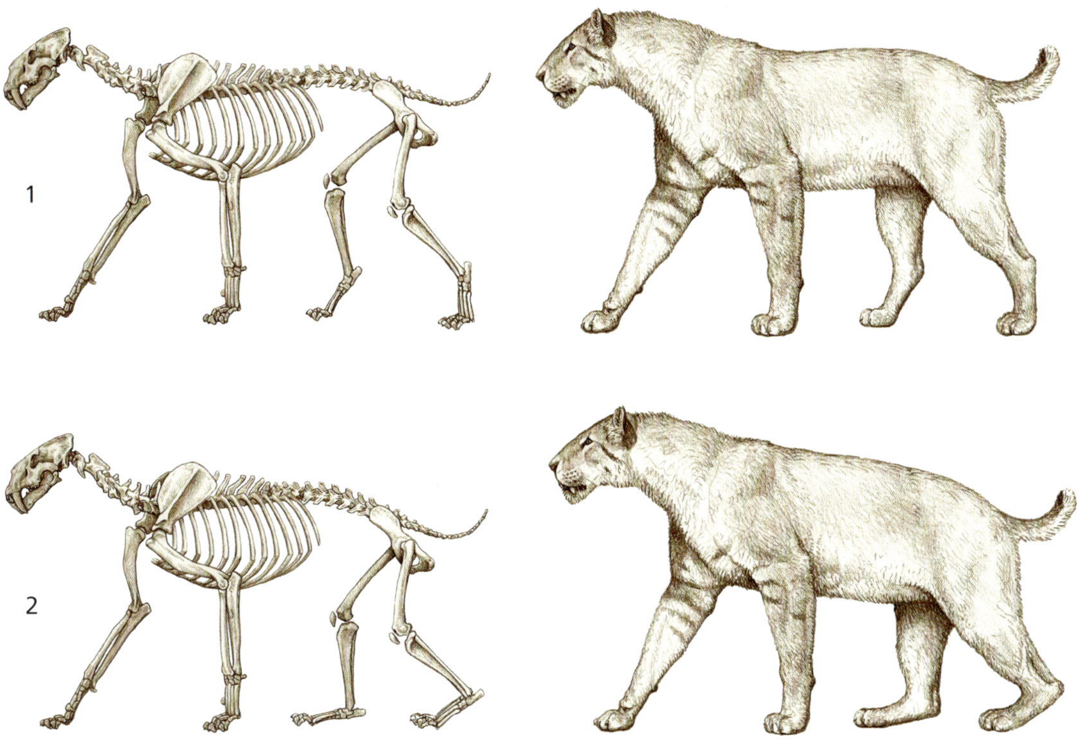

Fig. 5 Alternative reconstructions showing *Homotherium* as digitigrade and plantigrade: **1** digitigrade skeleton (left) and life appearance (right). – **2** plantigrade skeleton (left) and life appearance (right). Not only the morphology of the hind foot bones clearly indicates a digitigrade posture, but as the drawings show, the comparatively long (and undoubtedly digitigrade) forelimbs make it even more unlikely that the animal would walk on plantigrade hind limbs, since the steps of the hind limbs would become far shorter than those of the forelimbs. – (Artwork M. Antón).

Fig. 6 Reconstruction of a galloping *Homotherium latidens*. The animal's body proportions fit well with this kind of gallop or »canter« which is very efficient for covering long distances at moderate speeds. – (Artwork M. Antón).

We know that modern hypercarnivores above a certain body size will tend to hunt prey as large or larger than themselves, for reasons of energy balance (Sunquist/Sunquist 1989). But risk of injury is a controlling factor, and that is one reason why solitary hunters take relatively small prey. For instance, the leopard takes prey averaging 25 kg, which is below the predator's own body size (Hayward et al. 2006), but it can and will take prey larger than that, and often heavier than the predator (**fig. 8**).

Prey size in solitary cats largely depends on the body mass of the individual cat and on its adaptations for handling and controlling that prey. Thus, cheetahs, although overlapping in body weight with leopards, concentrate on gazelle-sized prey and smaller (**fig. 9**), due in part to the reduced grasping ability of their forelimbs.

Prey size in social cats, on the other hand, is not limited by individual power. Recent observations are showing that lions take very large prey more often than usually thought (Power/Compion 2009), and have precise techniques for doing so (**fig. 10**). Such access to large prey species is a function of the lion's group hunting capability, no matter what hypothesis we accept to explain the evolution of such group life.

Also, we know that carnivores with lesser strength and flexibility in their forepaws must resort to group action when it comes to taking larger prey, as they are ill-equipped to take such prey individually. This happens of course to modern dogs and hyaenas, and it may have happened with any extinct large predator that has sacrificed grasping power for speed and endurance in locomotion.

Dogs and hyaenas share with *Homotherium* the possession of large incisors, arranged in an arc (**fig. 11**). This feature is related with the greater role of the

Fig. 7 A sequence of drawings showing the succesive steps of the hypothetical killing bite in *Homotherium*: **1** mandible wide open, upper and lower canine tips in contact with the body of prey, but given the huge gape the muscles of the mandible cannot close it strongly enough. – **2** the whole head rotates down relative to the first neck vertebrae thanks to the pull of powerful neck muscles, and the upper canine pierces the prey's skin and flesh. – **3** Once the upper canines sink, the mandibular gape is reduced, so now the muscles of the mandible can provide strength to further close the jaws and deepen the bite. – (Artwork M. Antón).

Fig. 8 Photograph of a female leopard with its impala kill in the Samburu National Reserve, Kenya. Although the average weight of the leopard's prey is said to be around 25kg, these cats often hunt larger prey, and in this case the antelope is as heavy or heavier than the predator itself. – (Photo M. Antón).

Fig. 9 Video frame of a female cheetah killing a young gazelle in the Ngorongoro Conservation Area, Tanzania. Although cheetahs overlap in weight with leopards, they rarely take prey heavier than themselves and can be seen as essentially »gazelle specialists«. The adaptations of the cheetah for speed imply a partial loss of the grasping ability of its forelimbs, something that contributes to limit the size of the prey it can hunt. – (Filmed by M. Antón).

Fig. 10 Photograph of lionesses feasting on their giraffe kill in the Chobe National Park, Botswana. Although lions are the largest cats in Africa, hunting such enormous prey would be impossible for them without group action. – (Photo M. Antón).

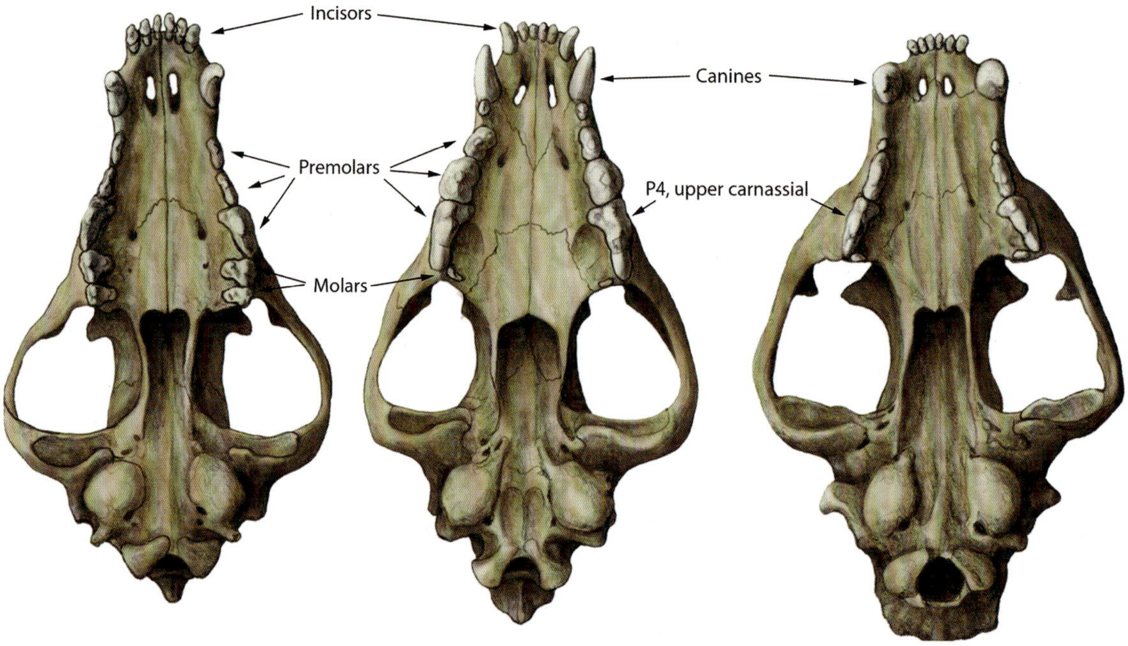

Fig. 11 Comparison between the skulls of three carnivore species in palatal (ventral) view, showing the differences in their dentitions. From left to right: wolf (*Canis lupus*), spotted hyaena (*Crocuta crocuta*) and leopard (*Panthera pardus*). Notice how the incisors of the wolf and hyaena are porpotionally larger and arranged in an arc, while those of the leopard are smaller and arranged in a row. This difference reflects the fact that in dogs and hyaenas the incisors play an important role in prey capture, while in felids that function is largely taken by the forelimbs. *Homotherium* had large, projecting incisors, suggesting some role in the capture of prey. – (Artwork M. Antón).

Fig. 12 Photograph of a group of cheetahs chasing zebra in the Maasai Mara Reserve, Kenya. The third member of the cheetah coalition was involved in the chase but is out of the frame. The targeted zebra foal was large enough to be outside the prey size range for solitary cheetahs, but coalitions do take such large prey. – (Photo M. Antón).

incisors for grasping prey, as compared to the felines where the grasping function is mostly taken by the forelimbs.

As we have seen, there are some similarities between the limb skeletons of *Homotherium* and the cheetah, especially in the shape and function of the wrist and forefeet. But these do not imply that the former was a sprint hunter, because other features of its anatomy, such as the comparatively short back and hind limbs, are poorly suited for quick acceleration or extreme top speeds, and fit better with efficient running at moderate speeds for relatively long distances (Antón et al. 2005). And yet, we may still learn from those similarities. The cheetah's physique is perfectly suited for taking medium-sized prey (gazelles) through solitary hunting. But cheetahs have a social system where male siblings often remain together through adult life, forming coalitions. Those coalitions are seen as an adaptation for the defence of territory and for easier access to transient females. However, during a trip to Kenya, I was able to film a coalition of three cheetah males attempting to hunt a large zebra foal, a bigger prey than any solitary cheetah could hope to catch (**fig. 12**). The cheetahs managed to bring down the foal, but it was rescued by a protective adult zebra in the last moment. Such observations may seem anecdotal, but they clearly show that even if group life is not originally and adaptation for the hunting of larger prey, it allows such predation when opportunities arise.

Fig. 13 Hypothetical scene set in the Pleistocene of Spain, showing a pair of *Homotherium* about to capture a large bovine. Group action would make it far easier to capture and control such large and powerful prey. – (Artwork M. Antón).

GROUP LIVING AND GROUP HUNTING

Do we have any reason to suspect that *Homotherium* would have a social structure independently of its hunting needs? In fact, studies of modern predators indicate that any large mammalian land carnivore living in relatively open environments, which it shares with an inflated large carnivore guild that includes lions and giant hyaenas, will be strongly pressed to become social in order to defend its resources such as territory, cubs, females and kills (Packer et al. 1990; Sunquist/Sunquist 1989). It may not react to all of these pressures, but if it reacts to just one of them, it will become social. And then a whole range of prey will become available: large bovids, for instance, can be taken (**fig. 13**), and the same happens with juvenile proboscideans (**fig. 14**).

Taking the young of large, aggressive herbivores is made easier by distracting maneuvers, something that group hunters such as modern wolves often do (**fig. 15**). *Homotherium* would obviously benefit from such strategies, a possibility that was poetically elaborated by Kurtén in his novel »Dance of the Tiger«.

The cheetah relies on its superb efficiency to make new kills if kleptoparasites rob from it, and its superior speed allows it to flee from any competing predator that could kill or maul it. So its strategy is to keep a low profile. But in the case of *Homotherium*, large body mass and moderate speed made those strategies less likely. In the face of aggression from competitors, it needed to stand its ground, and the best way to do this would be through strength by numbers.

So we see a number of reasons why it would be advantageous for *Homotherium* to live and hunt in groups, although obviously there is a large component of speculation.

Fig. 14 Hypothetical scene set in the Pleistocene of Spain, showing a group of *Homotherium* in the process of killing a juvenile mammoth. Again, group action would be required to bring down such prey, but also in order to keep other herd members at bay. – (Artwork M. Antón).

Fig. 15 Hypothetical scene set in the Pleistocene of Spain, showing an individual *Homotherium latidens* trying to capture a juvenile hippopotamus while a second cat distracts the mother. – (Artwork M. Antón).

SABERTOOTH CAT ECOMORPHS: SABERTOOTH DIVERSITY REVISITED

After looking at all those aspects of sabretooth functional anatomy and behaviour, we can look again at the diversity of body sizes and proportions among the sabretooth species from the Plio-Pleistocene of Europe. *Dinofelis* for instance could be compared with the modern jaguar in terms of body size and proportions, which could also imply a preference for wooded habitats. Moderate sabretooth adaptations might imply taking marginally larger prey than in the case of the modern jaguar. Its heyday in Europe coincides with a great extension of tropical and subtropical forests during the Early and Middle Pliocene. During the Late Pliocene there is evidence of a greater mosaicism, and during that time *Dinofelis* disappears, giving way to *Megantereon* and *Homotherium* during and after the Villafranchian.

Megantereon overlapped with the leopard in terms of body size although some specimens, such as the Senèze skeleton, are larger than any leopard, and are more like jaguars. Body proportions suggest a closed habitat, as in the case of *Dinofelis*, but more extreme sabretooth adaptations in the neck and skull suggest larger prey relative to body size. Smaller size than *Dinofelis* could help *Megantereon* to partition resources with the larger *Homotherium*. Frequent coexistence of *Megantereon* and *Homotherium* at Villafranchian fossil sites points to a mosaic of vegetation at those areas, and at their ability to hunt different prey at different times of day and/or in different sections of the habitat.

Homotherium, in turn, overlapped with the lion in body size and our own comparisons of post-cranial, and especially appendicular morphology, suggest that it was a relatively light animal, certainly lighter than the large Pleistocene lions that shared its habitat. Marked sabretooth specializations allowed the taking of large prey, while adaptations for long-distance running limited their prey-handling abilities, but allowed the animal to range across large territories and maybe even to follow migrating prey for part of their seasonal trips.

CONCLUSIONS

In conclusion, although broad parallels can be established with modern felines, sabretooth species were no exact ecological equivalents of jaguars, leopards or lions, as further demonstrated by the fact that they coexisted with those species at various points in space and time. They were unique products of evolution with their own answers to the challenges of hunting large prey.

Homotherium itself was a broadly lion-sized sabretooth cat, which combined a craniodental and cervical complex specialized for the killing of large prey, with relatively long limbs adapted for running at moderate speeds for long distances. Such locomotor adaptations implied reduced individual power for subduing prey. *Homotherium* likely used the open sections of its habitat and foraged widely, probably following migratory prey. In order to catch prey of the right size for its killing weapons it may have needed to hunt in groups. As a large felid living in open habitats with »crowded« large carnivore guilds, it had additional reasons to become social: protection of territory, of females, of cubs and of prey.

REFERENCES

Antón 2013: M. Antón, Sabertooth (Bloomington, Indianapolis 2013).

Antón/Galobart 1999: M. Antón / A. Galobart, Neck Function and Predatory Behaviour in the Scimitar Toothed Cat *Homotherium latidens* (Owen). Journal of Vertebrate Palaeontology 19/4, 1999, 771-784.

Antón/Sánchez 2004: M. Antón / I. M. Sánchez, Art and science: the methodology and relevance of the reconstruction of fossil vertebrates. In: E. Baquedano Pérez / S. Rubio Jara (eds), Miscelánea en homenaje a Emiliano Aguirre. 2: Paleontología. Zona Arqueológica 4 (Madrid 2004) 74-94.

Antón/Galobart/Turner 2005: M. Antón / A. Galobart / A. Turner, Co-existence of Scimitar-toothed Cats, Lions and Hominins in the European Pleistocene. Implications of the Post-cranial Anatomy of *Homotherium latidens* (Owen) for Comparative Palaeoecology. Quaternary Science Reviews 24, 2005, 1287-1301.

Antón/García-Perea/Turner 1998: M. Antón / R. García-Perea / A. Turner, Reconstructed facial appearance of the sabretoothed felid *Smilodon*. Zoological Journal of the Linnean Society 124, 1998, 369-386.

Antón et al. 2004: M. Antón / M. J. Salesa / J. F. Pastor / I. M. Sánchez / S. Fraile / J. Morales, Implications of the Mastoid Anatomy of Larger Extant Felids for the Evolution and Predatory Behaviour of Sabre-toothed Cats (Mammalia, Carnivora, Felidae). Zoological Journal of the Linnean Society 140, 2004, 207-221.

2009: M. Antón / M. J. Salesa / A. Galobart / J. F. Pastor / A. Turner, Soft tissue reconstruction of *Homotherium latidens* (Mammalia, Carnivora, Felidae). Implications for the possibility of representation in Palaeolithic art. Geobios 42, 2009, 541-551.

Ballesio 1963: R. Ballesio, Monographie d'un Machairodus du gisement villafranchien de Senéze: *Homotherium crenatidens* fabrini. Travaux du Laboratoire de Géologie de la Faculté de Sciences de Lyon N. S. 9 (Saint-Étienne 1963).

Bryant/Russell 1992: H. N. Bryant / P. Russell, The role of phylogenetic analysis in the inference of unpreserved attributes of extinct taxa. Philosophical Transactions of the Royal Society of London B 337, 1992, 405-418.

Freudenberg 1914: W. Freudenberg, Die Säugetiere des älteren Quartärs von Mitteleuropa. Geologische und Paläontologische Abhandlungen N. F. 12, 1914, 455-671.

Hayward et al. 2006: M. W. Hayward / P. Henschel / J. O'Brien / M. Hofmeyr / G. Balme / G. I. H. Kerley, Prey preferences of the leopard (*Panthera pardus*). Journal of Zoology 270, 2006, 298-313.

Packer/Scheel/Pusey 1990: C. Packer / D. Scheel / A. Pusey, Why lions form groups: food is not enough. American Naturalist 136, 1990, 1-19.

Power/Compoin 2009: R. J. Power / R. X. S. Compion, Lion Predation on Elephants in the Savuti, Chobe National Park, Botswana. African Zoology 44/1, 2009, 36-44.

Schaub 1925: S. Schaub, Über die Osteologie von *Machairodus cultridens* Cuvier. Eclogae Geologica Helvetiae 19, 1925, 255-266.

Sunquist/Sunquist 1989: M. E. Sunquist / F. C. Sunquist, Ecological constrains on predation by large felids. In: J. L. Gittleman (ed.), Carnivore Behavior, Ecology and Evolution (London 1989) 220-245.

Turner/Antón 1999: A. Turner / M. Antón, Climate and evolution: implications of some extinction patterns in African and European machairodontine cats of the Plio-Pleistocene. Estudios Geologicos 54, 1999, 209-230.

Witmer 1995: L. M. Witmer, The extant phylogenetic bracket and the importance of reconstructing soft tissues in fossils. In: J. J. Thomason (ed.), Functional Morphology in Vertebrate Paleontology (Cambridge 1995) 19-33.

SUMMARY / ZUSAMMENFASSUNG

Behaviour of *Homotherium* in the Light of Modern African Big Cats

In order to reconstruct the appearance and behaviour of the sabretoothed felid *Homotherium* we combine the study of its functional anatomy with various complementary sources of information, including the comparative anatomy, the dissection of extant carnivores and the observation of living felines in the African wilderness. The results of these comparative studies reveal *Homotherium* as a lion-sized sabretooth, much lighter in build than its Pleistocene relatives such as the *Smilodon*. Compared to extant and fossil lions, *Homotherium* had a longer and more muscular neck; a shorter back and tail; narrower forearms and less powerful forepaws with smaller and less retractable claws, with the exception of a huge dewclaw. These features suggest that *Homotherium* would target large ungulates as prey, but in order to bring down animals the size of a buffalo and larger, the loss of strength and of the prehensile abilities of its forepaws would pose a limitation, and make group action necessary. Body proportions and morphology suggest an adaptation to relatively open environments, and given this and the cat's body size there could be ecological pressures to evolve group living, in order to defend territory, cubs, and kills. In turn, we know from the observation of living felines that group-living, even if originally developed as a response to ecological pressures, increases the likelihood of group hunting.

Das Verhalten von *Homotherium* im Lichte der modernen afrikanischen Großkatzen

Um das Aussehen und das Verhalten des Säbelzahntigers *Homotherium* zu rekonstruieren, kombinieren wir die Untersuchung seiner funktionellen Anatomie mit verschiedenen ergänzenden Informationsquellen, darunter die vergleichende Anatomie, die Sektion lebender Raubtiere und die Beobachtung lebender Raubkatzen in der afrikanischen Wildnis. Die Ergebnisse dieser vergleichenden Studien zeigen, dass *Homotherium* eine löwengroße Säbelzahnkatze ist, die viel leichter gebaut ist als ihre pleistozänen Verwandten wie der Smilodon. Im Vergleich zu heutigen und fossilen Löwen hatte *Homotherium* einen längeren und muskulöseren Hals, einen kürzeren Rücken und Schwanz, schmalere Unterarme und weniger kräftige Vorderpfoten mit kleineren und weniger einziehbaren Krallen, mit Ausnahme einer großen Afterkralle. Diese Merkmale deuten darauf hin, dass *Homotherium* große Huftiere als Beute ins Visier nahm, aber um Tiere von der Größe eines Büffels und größer zu erlegen, würde der Verlust der Kraft und der Greiffähigkeit der Vorderpfoten eine Einschränkung darstellen und Gruppenaktionen erforderlich machen. Die Körperproportionen und die Morphologie deuten auf eine Anpassung an relativ offene Umgebungen hin, und angesichts dessen und der Körpergröße der Katze könnte ein ökologischer Druck bestehen, ein Gruppenleben zu entwickeln, um Territorium, Jungtiere und Beute zu verteidigen. Im Gegenzug wissen wir aus der Beobachtung lebender Katzen, dass das Leben in der Gruppe, selbst wenn es sich ursprünglich als Reaktion auf ökologischen Druck entwickelt hat, die Wahrscheinlichkeit der Gruppenjagd erhöht.

DICK MOL · BRAM W. LANGEVELD

ON REMAINS OF *HOMOTHERIUM* FROM THE BOTTOM OF THE NORTH SEA BETWEEN THE BRITISH ISLES AND THE NETHERLANDS AND THEIR STRATIGRAPHICAL AGE

The Southern Bight of the North Sea between the British Isles and the Netherlands is a rich source of fossil mammal remains dating from the entire Pleistocene (e.g., Kortenbout van der Sluijs 1971a; 1971b; 1983; 1985; van Kolfschoten/Laban 1995; Mol et al. 2003; 2006; 2013; Reumer et al. 2003; Mol/Reumer 2010). Some of these specimens are rare and very important. Fishermen have been collecting fossils from their nets since at least 1874 (**figs 1-2**; Mol et al. 2008a). Initially, localities near the Brown Bank were recognised as being prolific sources of finds, but later, many other fossil-rich sites in the Southern Bight were discovered. Current knowledge of these fossil localities is the result of intense collaboration between museums, private collectors and the fishing industry. It all began when the curator of fossil mammals of the former National Museum of Geology and Mineralogy, G. Kortenbout van der Sluijs, conducted regular visits to Dutch fishing ports in the 1960s to obtain fossil bones for his museum. He found that the larger vessels (>30m in length) that went for flatfish outside the 12-mile zone were particularly successful in collecting fossil mammal remains as a bycatch of their haul of fish.

Fig. 1 A fishing vessel returning to the harbour at Stellendam (prov. Zuid-Holland/NL) after a 24-hour-trip, with a large amount of Late Pleistocene mammal remains on board. – (Photo H. Wildschut).

In: Nicholas J. Conard · Henning Hassmann · Kurt Felix Hillgruber · Jordi Serangeli · Thomas Terberger (eds), The *Homotherium* Finds from Schöningen 13II-4. Man and Big Cats of the Ice Age. Forschungen zur Urgeschichte aus dem Tagebau von Schöningen 4 (Mainz 2022). DOI: https://doi.org/10.11588/propylaeum.1006.c13520

Fig. 2 Large male woolly mammoth skull trawled from the seabed at the quay in the harbour of Stellendam. – (Photo F. van der Vossen).

Subsequently, it appeared that smaller vessels that fished within the 12-mile zone also brought many skeletal remains back to port. These vessels are the so-called Eurocutters with engines of less than 300 bhp and a length of <22 m. Fossils collected by these smaller ships originated from an area north of the mouth of the Western Scheldt estuary, offshore from the provinces of Zeeland and Zuid-Holland. In the early 1990s, Eurocutters also started fishing in the so-called Eurogeul, the 28-m-deep shipping lane that connects with Rotterdam harbour (prov. Zuid-Holland/NL). This now appears to be the most productive locality, together with the region to the south of it. This latter site has become a major source for dredging sand in order to construct the new Rotterdam harbour extension, the so-called Maasvlakte 2. In 2001, we began a large-scale operation to fish for fossils specifically for research purposes (Mol/Post 2010); this work includes searching in dredged sand on beaches (Mol et al. 2013). As a result of all these activities, we now have a wealth of material from the area of the North Sea that has become known as the Southern Bight.
All material obtained from the bottom of the North Sea through fishing (as well as sand dredging) is obviously obtained *ex situ*. Therefore, specimens cannot be dated on the basis of their stratigraphical context. The lack of *in situ* context might seem to be a considerable challenge to their study, but over the past few decades we have observed that this is not the case. Based on the known stratigraphical ranges of species from *in situ* occurrences, combined with the state of preservation of the North Sea fossils and knowledge of the geology of the sea floor, it has proved possible to determine their age unequivocally. Furthermore, radiocarbon dating provides exact ages to fossils from <45 kyr BP. Finally, the question regarding exactly which stratigraphical data has been lost through the recovery *ex situ* should be addressed. The Southern Bight of the North Sea has witnessed repeated exposure and submersion of the shelf during the Pleistocene, especially during the Late Pleistocene, when it was exposed and inundated again within roughly 100 000

Fig. 3 Early and Late Pleistocene mammoth remains can be easily identified based on their morphology: **1** heavily mineralised, dark coloured upper M3 of *Mammuthus meridionalis*. – **2** barely mineralised upper M3 of *Mammuthus primigenius*. – (Photo H. Wildschut).

years. In addition, the character of the rivers running through it, notably the River Rhine-Meuse (Laban/Rijsdijk 2002), has changed repeatedly through Pleistocene climatic fluctuations, changing from a braided river system into a meandering one and vice versa (Berendsen 2011). All of this has resulted in the strong reworking of the majority of fossiliferous deposits, as we have observed on the basis of borehole cores taken from the Eurogeul area. For this reason, data obtained from the prime stratigraphical context of the fossils is limited anyway.

It is a fact that North Sea fossils of species typical of the Early or early Middle Pleistocene invariably are heavily mineralised and dark brown to black in colour. They produce a high-pitched/metallic sound when tapped upon with a solid object. Furthermore, they often contain significant amounts of iron, which may render a rusty appearance in some small parts. Finally, these fossils are normally obtained from a limited number of sites in the North Sea. Easily distinguished from this is the much more common type of preservation from the North Sea which is invariably observed in species typical of the Late Pleistocene or Early Holocene (**fig. 3**). These fossils are barely mineralised at all and are still very suitable for radiocarbon dating (Mol et al. 2006) and DNA analysis (Alter et al. 2015). They are often lighter in colour than the older material. These fossils are retrieved in great abundance from an extensive number of sites in the North Sea.

Based on the above observations, it is possible to assign with confidence a rough stratigraphical date to specimens that cannot be aged by morphology alone. For instance, a metacarpal of a large horse, *Equus* sp., can be confidently assigned either an Early to early Middle Pleistocene or a Late Pleistocene age by direct comparison with the preservation of other material obtained from the same area. We cannot stress enough the clear differences between the older and the younger material, which have proved so useful to our interpretations over the past few decades. In fact, this is a crucial point concerning the Late Pleistocene mandible of the sabre-toothed cat described below.

SITES AND FAUNAS

The Southern Bight of the North Sea, especially areas such as the Brown Bank and the Eurogeul, has produced large quantities of skeletal elements from the entire Pleistocene and Early Holocene (Mol et al. 2006). Based on radiocarbon dates, the Late Pleistocene fauna includes, amongst others: woolly mammoth / *Mammuthus primigenius* (Blumenbach, 1799); woolly rhinoceros / *Coelodonta antiquitatis* (Blumenbach,

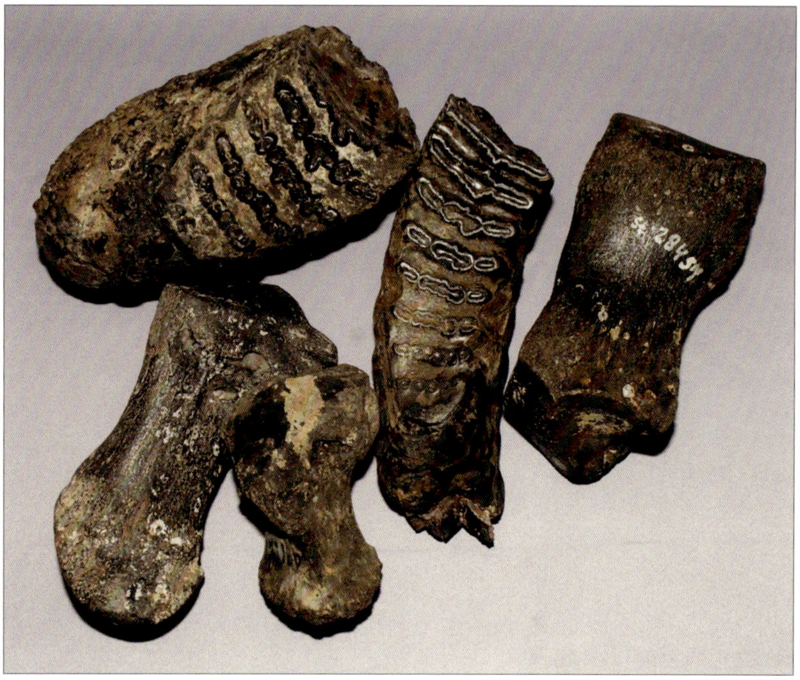

Fig. 4 Early Pleistocene *Mammuthus meridionalis* remains from the Eastern Scheldt estuary and of the so-called Deep Water Channel in the North Sea. – (Photo H. Wildschut).

1799); wild horse / *Equus caballus* Linnaeus, 1758; giant deer / *Megaloceros giganteus* (Blumenbach, 1799); reindeer / *Rangifer tarandus* (Linnaeus, 1758); steppe bison / *Bison priscus* (Bojanus, 1825); cave hyena / *Crocuta spelaea* (Goldfuss, 1823) and muskox / *Ovibos moschatus* (Zimmermann, 1780). Amongst other taxa, the Early Holocene fauna comprises red deer / *Cervus elaphus* Linnaeus, 1758; roe deer / *Capreolus capreolus* (Linnaeus, 1758); moose / *Alces alces* (Linnaeus, 1758); wild boar / *Sus scrofa* Linnaeus, 1758; otter / *Lutra lutra* Linnaeus, 1758 and humans / *Homo sapiens* Linnaeus, 1758. There are numerous sites that yield these Late Pleistocene and Early Holocene faunas in abundance. However, localities with Early to early Middle Pleistocene material are much rarer, and the quantity of these fossils retrieved per site is much lower. These localities are, therefore, of great importance, particularly in order to understand sabre-toothed cats.

Homotherium cf. *latidens* (Owen, 1846)

One of the localities that has produced sabre-toothed cat remains is an estuary of the River Scheldt in the province of Zeeland (Hooijer 1962; Mol et al. 2008b). Fauna from the Early Pleistocene, recently redated between 2.5 and 2 million years old, by biostratigraphy (Scager 2015), has been recovered from the Eastern Scheldt. This assemblage includes, amongst others, the mastodon of Auvergne / *Anancus arvernensis* (Croizet/Jobert, 1828), southern mammoth / *Mammuthus meridionalis* (Nesti, 1825) (**fig. 4**), Perrier's hyena / *Pliocrocuta perrieri* (Croizet/Jobert, 1828) and rare *Homotherium*. D. A. Hooijer (1962) was the first to note a mandible fragment which he attributed to *Homotherium* cf. *latidens*. This incomplete mandible was collected from the Eastern Scheldt seabed by the mussel cutter ZZ8 of the Schot brothers of Zierikzee (the Netherlands). D. A. Hooijer (1962) described the specimen as dark brown and moderately mineralised, substantially less so than the Early Pleistocene remains that are usually collected in the deep gullies of the Eastern Scheldt. He thus assumed the deposits in which the specimen was embedded to have been of Middle Pleistocene age, i.e., between 800000 and 200000 years old. However, we have compared the 1962 specimen with remains of *Anancus arvernensis* from the same site. These have the same degree of mineralisation and the same dark brown colours as the *Homotherium* specimen described by D. A. Hooijer (1962).

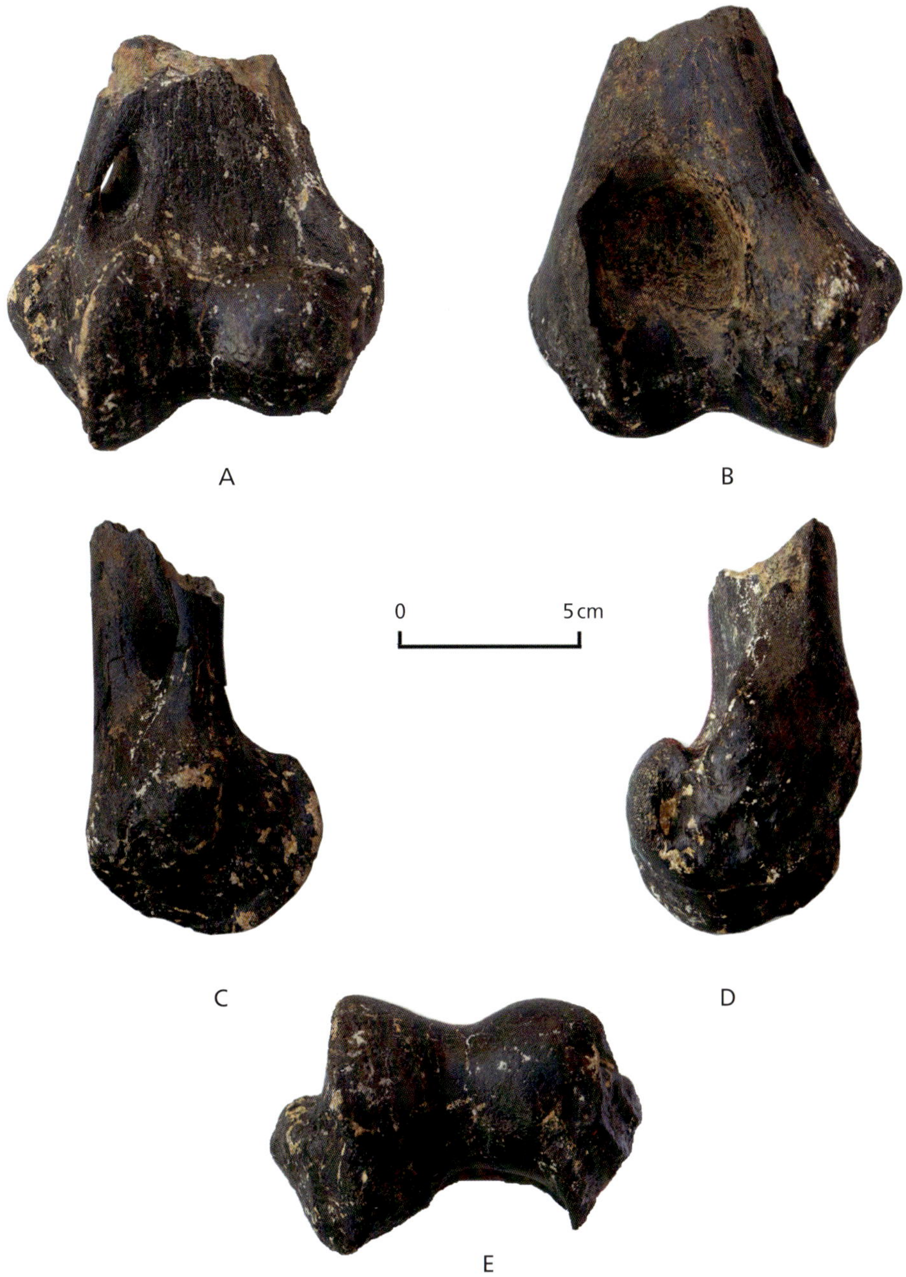

Fig. 5 The distal end of the left humerus of *Homotherium crenatidens* from the North Sea: **A** anterior view. – **B** posterior view. – **C** medial view. – **D** lateral view. – **E** distal view of the trochlea. – (Photos H. Wildschut).

Since there are no Middle Pleistocene sediments known from the studied Eastern Scheldt area, and since characteristic Middle Pleistocene faunal elements are missing from the local fossil record, we attribute this specimen to the Early Pleistocene assemblage.

Homotherium latidens (Owen, 1846)

In 1971, the mussel cutter ZZ8 collected another fossil of *Homotherium*: an incomplete right calcaneum from the Flauwerspolder locality in the Eastern Scheldt estuary. Again, this specimen shows features that are so typical of Early Pleistocene faunal elements from the Eastern Scheldt such as *Anancus arvernensis* and *Mammuthus meridionalis*. Both the 1962 and 1971 specimens are housed in the collections of Naturalis Biodiversity Center (Leiden).

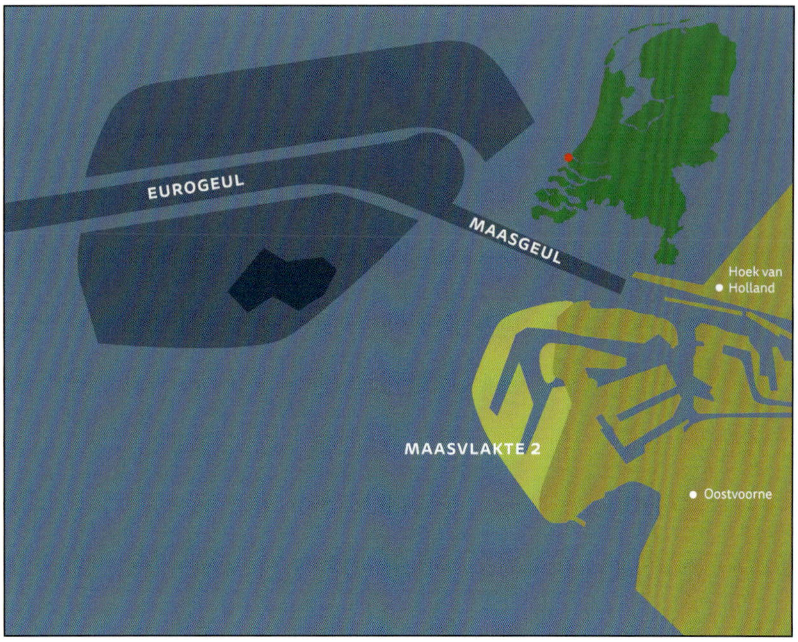

Fig. 6 Schematic view of Maasvlakte 2 and the Eurogeul area, including the sand source area off the coast of the province of Zuid-Holland. Inset: location of Maasvlakte 2 (red dot) in the Netherlands. – (Map J. van Leeuwen / J. Streutker).

More remains that are attributable to *Homotherium latidens* were reported by van Hooijdonk (1999) and Mol et al. (2008b), including a right calcaneum from the locality »Onrust«. This site is located in the North Sea off the coast of Walcheren (prov. Zeeland/NL). From fairly close to »Onrust«, in the »Roompot« area (Eastern Scheldt), there are two metacarpals of *Homotherium latidens*. Again, these remains are heavily mineralised and characterised by a brown/blackish hue. Therefore, they are assigned an Early Pleistocene age. The specimens from both »Onrust« and »Roompot« are in the private collection of Kees van Hooijdonk (Rucphen, the Netherlands).

From a locality named »Het Gat« in the North Sea, east of the so-called Brown Bank, the trawler GO41 from Stellendam collected a distal portion of a humerus, which was attributed to *Homotherium latidens* by Mol et al. (2003). This well-preserved specimen is thoroughly black in colour and extremely heavily mineralised. When tapped upon with a hard object, the fossil produces a high-pitched sound. The locality »Het Gat« is famous for a fauna which is dated at c. 1 Ma, including an advanced type of *Mammuthus meridionalis* or an early type of the steppe mammoth / *Mammuthus trogontherii* (Pohlig, 1885), plus the extinct hippopotamus / *Hippopotamus antiquus* Desmarest, 1822. The specimen is housed in the Dick Mol Collection (Hoofddorp, the Netherlands).

Homotherium crenatidens (Weithofer, 1889)

Van Logchem/Mol (2008) described and illustrated (**fig. 5**), for the first time from the bottom of the North Sea off the coast of East Anglia (United Kingdom), a distal end of a left humerus of an Early Pleistocene *Homotherium crenatidens* (Weithofer, 1889). This humerus was trawled from the seabed (52° 50′ N, 02° 18′ E) by the fishing vessel TX1, the Klasina-J. of skipper Cor Vonk of Texel (the Netherlands), in August 2008. The specimen, heavily mineralised and thoroughly black, was identified by van Logchem/Mol (2008) as *Homotherium crenatidens* on the basis of its large proportions in comparison with humeri of other felids such as *Homotherium latidens*. *Homotherium crenatidens* is a heavily built sabre-toothed cat and, as such, also referred to in the literature as the »greater scimitar cat«. The weight of this species could be as much as 400 kg (Hemmer 2001, 2004). It is considered the direct ancestor of *Homotherium latidens* (Ballesio 1963).

Fig. 7 Trailing suction hopper dredger creating new land (Maasvlakte 2) using Pleistocene sediments, including fossils, from the Eurogeul area. – (Photo H. Wildschut).

The trawling area off the coast of East Anglia is well known for remains of Early Pleistocene faunal elements such as *Mammuthus meridionalis*, the Etruscan rhinoceros / *Stephanorhinus etruscus* (Falconer, 1868) and large deer / *Eucladoceros ctenoides* (Nesti, 1841). All remains of these species, dated as Early Pleistocene, are characterised by their dark brownish to black colours, matching the *Homotherium* specimen. *Homotherium crenatidens* is considered a carnivorous member of such an Early Pleistocene faunal association. When tapped upon with a hard object, these fossils produce a metallic sound. The specimen belongs to the private collection of Bert Schagen of Texel (the Netherlands).

MAASVLAKTE 2 YIELDS NEW *HOMOTHERIUM* RECORDS

In addition to summarising previous records of *Homotherium* from the North Sea, we also present four new records here. These were all recovered by private collectors from the beach of Rotterdam's youngest harbour extension, i.e., Maasvlakte 2. First, we discuss Maasvlakte 2 and the state of palaeontological research of this young site; we then describe and illustrate the new specimens.

The area of the Maasvlakte 2 comprises 240 million m³ of sediment, which was sucker-dredged from the Eurogeul area by trailing suction hopper dredgers (**figs 6-7**). This sediment was carefully deposited seawards of the previous Rotterdam harbour extension (Maasvlakte, now often referred to as the first Maasvlakte, or Maasvlakte 1) in order to obtain a significant area for new harbour extensions. Its main goal is enabling the harbour industry to maintain and improve the position of Rotterdam harbour as a globally important port. The engineers responsible for retrieving the vast volume of sediment from the Eurogeul area needed for the construction came up with two options. Option 1 was a more or less classic approach, where sediment

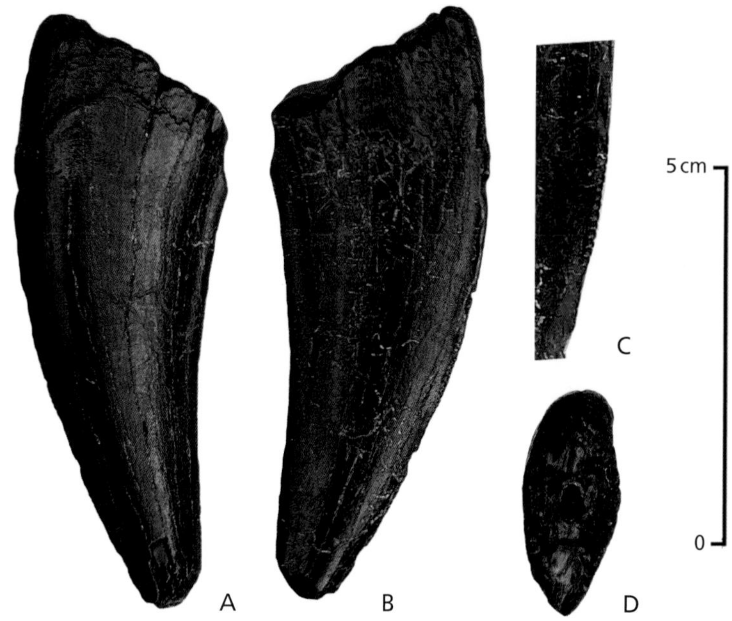

Fig. 8 *Homotherium latidens*, crown of an upper left canine (C sup. sin.) from Maasvlakte 2: **A** exterior view. – **B** interior view. – **C** detail of the enamel showing the typical serration. – **D** cross section of the root. – (Photos A-C H. Wildschut; D M. Simmelink).

would be dredged up to 4-6 m below the original seabed. This would obviously require a vast area in order to obtain the necessary volume. The other option was a bolder one: retrieve all sediment needed from only a relatively small area by dredging not 6 m below the original seabed, but down to a depth of up to 20 m. Even though recovery of the sea floor dredged to 20 m depth would take significantly longer, the lesser extent of the area disturbed made it the better alternative environmentally speaking. That environmental impact was an important factor in the final decision to proceed with the second option. Palaeontologically speaking, this decision could not have been better.

Prior to the construction of Maasvlakte 2, the Eurogeul and surrounding area were known solely for their huge abundance of well-preserved Late Pleistocene and Early Holocene mammal fossils (Mol et al. 2006). However, the extensive deepening of part of the area for the construction of Maasvlakte 2 resulted in palaeontological gold: sediments from the late Middle Pleistocene and possibly Early Pleistocene reached for the first time in this area (Busschers/van Heteren/Westerhoff 2012) appeared to yield significant volumes of Early to early Middle Pleistocene (partially reworked) fossils, as noted almost directly after the beach was declared open to the general public. This was triggered by the intense collecting activities on the Maasvlakte 2 beach by a large number of private collectors who had been waiting impatiently for free access to the beach. Based on their collections and the wealth of data contained in them Mol et al. (2013; 2015) and Mol/Langeveld (2015a; 2015b) identified c. three faunal assemblages from Maasvlakte 2, including one dated as Early Pleistocene or early Middle Pleistocene that had not been recovered from the Eurogeul area previously. Furthermore, the careful collecting by the private collectors has resulted in significant additions to the already well-known Late Pleistocene and Early Holocene faunas, through recovery of small- to medium-sized vertebrate remains. These include skeletal elements of e. g., hares and diverse birds (Mol/Langeveld 2016).

For the composition of Late Pleistocene and Early Holocene faunas from Maasvlakte 2, reference is made to the literature on the Eurogeul area (e. g., Mol et al. 2006) and the above. Species recovered from the Maasvlakte 2 beach assigned an Early to early Middle Pleistocene date include: *Mammuthus meridionalis*; *Stephanorhinus etruscus*; large horse / *Equus bressanus* Viret, 1954; wild boar / *Sus strozzii* Forsyth Major, 1881; extinct moose / *Alces latifrons* (Johnson, 1874); desman / *Desmana thermalis* Kormos, 1930 and giant beaver / *Trogontherium cuvieri* Fischer, 1809 (Mol et al. 2013; Mol/Langeveld 2015a; 2015b). Possibly slightly younger are remains of forest rhinoceros / *Stephanorhinus kirchbergensis* (Jäger, 1839) and a hippopotamus / *Hippopotamus incognitus* Faure, 1984. Again, remains of species dated to the Early to early Middle Pleistocene all show the same style of preservation: dark brown to black in colour, often with some rusty sediment adhering, invariably thoroughly mineralised, and yielding a high-pitched sound when tapped upon with a hard object. The four new *Homotherium* records described below show the very same preservational style and are therefore confidently dated to this Early to early Middle Pleistocene fauna.

The first Maasvlakte 2 specimen of *Homotherium latidens* is a more or less complete crown of an upper left canine (C sup. sin.), found on the beach by Marc Simmelink, March 15, 2015 (**figs 8-9**). This canine fragment, with a maximum length of 75 mm, shows all characteristics that allow it to be attributed to *Homotherium latidens*. The crown is typically laterally flattened and the enamel is completely serrated, with 22 serrations/cm. The tip of the crown is damaged; this may have occurred *in vivo*. The interior side of the crown is more or less smooth, whereas the exterior side is rounded. The anterior-posterior length, measured at the base of the crown, is 28.5 mm; the maximum thickness, measured interiorly-exteriorly is 12 mm. The overall preservation of the specimen suggests it to have been reworked from older sediments. It reveals all previously described characters of Early/early Middle Pleistocene fossils from Maasvlakte 2. It is housed in the collection of Marc Simmelink (Hellevoetsluis, the Netherlands). A cast of this specimen is stored in the collection of the Natural History Museum Rotterdam (NMR999100012093).

The second Maasvlakte 2 specimen (**fig. 10**) is also a fragmentary upper canine (C. sup. dex.); it is better preserved. On

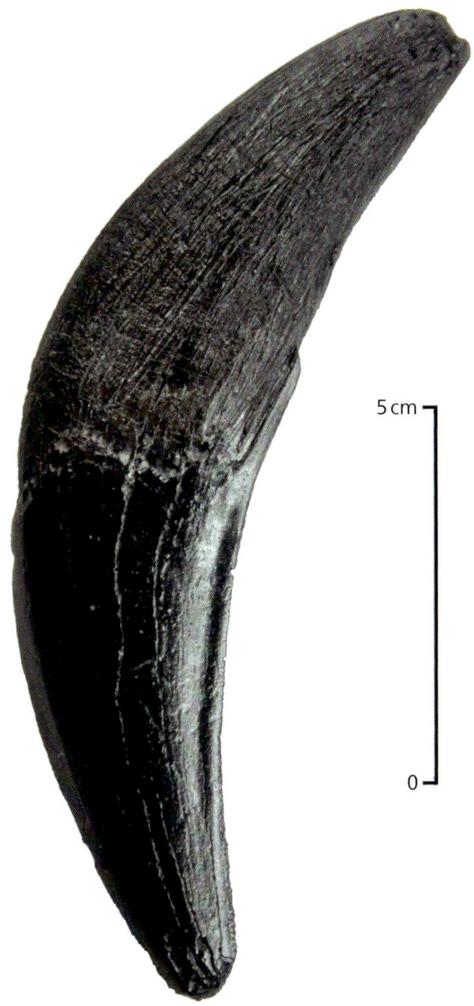

Fig. 9 Reconstructed upper left canine (C sup. sin.) of *Homotherium latidens* using a cast of the crown illustrated in **fig. 8**. – (Photo H. Wildschut).

Fig. 10 *Homotherium latidens*, crown of an upper right canine (C sup. dex.) from Maasvlakte 2: **A** exterior view. – **B** interior view. – **C** detail of the enamel showing the typical serration. – **D** cross section of the root. – (Photos B. W. Langeveld).

Fig. 11 *Homotherium latidens*, right lower canine (C inf. dex.) from Maasvlakte 2: **A** exterior view – **B** interior view – **C** posterior view. – (Photos B. W. Langeveld).

the basis of size it must have belonged to a slightly larger individual. It was collected on June 6, 2016 from the beach. This fossil shows part of the root, the base of the enamel and most of the crown, the tip having been broken off. The maximum length of the specimen is 71 mm and the serrations number 21 per cm. The anterior-posterior length, measured at the base of the crown, is 30 mm; the maximum thickness at the root, measured interiorly-exteriorly is 12 mm. The fractured surface at the upper end of the crown appears to be ancient damage. The preserved root is dark brown to black in colour and the pulp cavity contains the typical red rusty sediment. The colour of the enamel on both sites of the crown is olive green with some brown to black striations. When observed at slightly higher magnification, the serrations are almost translucent. This differs markedly from the first specimen. Although showing all previously described typical characters of Early/Middle Pleistocene fossils from Maasvlakte 2, it is differentiated from the first specimen in that it was not, apparently, reworked. The second specimen, catalogued as WL0893, is housed in the collection of its collector Walter Langendoen (Hellevoetsluis, the Netherlands). The collection of the Natural History Museum Rotterdam contains a cast (NMR999100012090) of this specimen.

The third new specimen (**fig. 11**), also collected by Walter Langendoen (WL0894, plus a cast in the Natural History Museum Rotterdam, NMR999100012091) in the spring of 2016, consists of a complete lower right canine (C inf. dex.), of which the tip of the crown is worn. The maximum height is 56 mm, the maximum width, at the base of the crown, 11 mm and the maximum width in the middle of the root, 12 mm. The anterior-posterior length of the crown is 17 mm and the maximum anterior-posterior length of the root is 19 mm. The maximum height of the crown is 20 mm. The root of the specimen is flattened and slightly S-shaped interior to exterior; the crown has a laterally flattened cross section. The specimen is heavily mineralised, and the enamel is of an olive green hue with some black. The enamel is worn on the anterior side. The base of the crown on the posterior side shows five serrations. Above that part the enamel is worn. All observed characters allow attribution to *Homotherium*. The preservation shows its Early to early Middle Pleistocene age.

The fourth and last new specimen presented here (**fig. 12**) is a fragment of a right mandible, which was collected by René Lamers on May 5, 2016 from the beach of Maasvlakte 2. It consists of the most rostral part of the jaw, including the symphysis, the alveoli of i2 and i3, the root of the canine still in place, the alveoli of the p3 and a tiny part of the rostral alveolus of the p4. Behind and below this alveolus the jaw is broken off. The maximum length of the fossil is 63 mm; the maximum height equals 54 mm. The maximum thickness of the symphysis is 26 mm. Although being just a small fragment without any discernible characteristics on the sole dental element preserved, it is undoubtedly *Homotherium.* Proof is found in the very typical morphology of the fragment (e.g., its slope and slenderness directly caudal of the canine, the tiny alveoli of the p3, the flat rostral side of the symphysis and the oval cross section of the root) in direct comparison with the complete North Sea mandible described below. The jawbone is heavily mineralised and produces a high-pitched sound when tapped upon with a hard object. The pulp cavity of the canine root shows some rusty sediment, as does the border between the root and its alveolus. The specimen appears to have been reworked. Therefore, it can be dated as Early to early Middle Pleistocene. It is kept in the collection of its col-

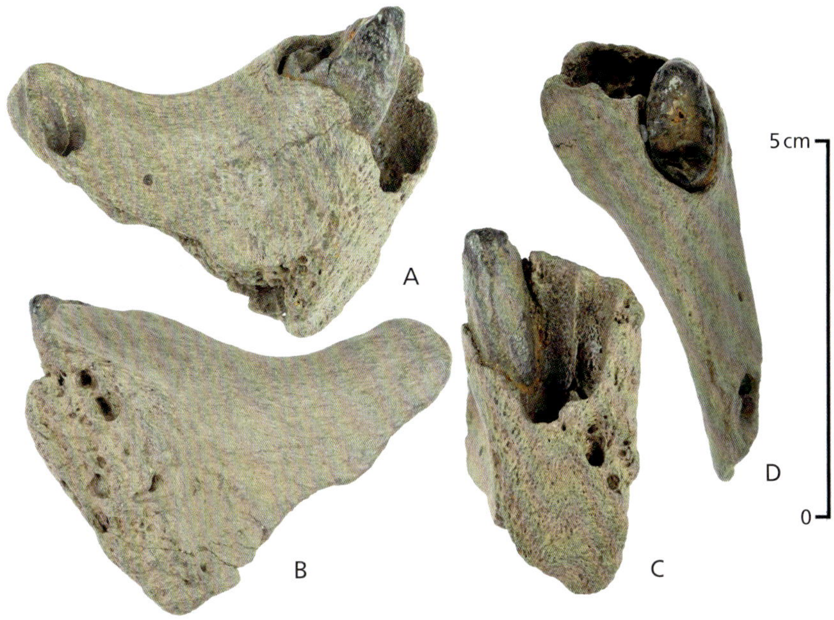

Fig. 12 *Homotherium latidens*, fragment of right mandible with remnant of the canine from Maasvlakte 2: **A** buccal view – **B** lingual view – **C** cranial view – **D** dorsal view. – (Photos B. W. Langeveld).

lector, René Lamers (Rotterdam, catalogued as 219). A cast is stored in the collection of the Natural History Museum Rotterdam (NMR999100012092).

LATE PLEISTOCENE *HOMOTHERIUM LATIDENS*

One of the most remarkable palaeontological finds made during beam-trawling activities in the Southern Bight of the North Sea is a very well-preserved half mandible of a large felid from a locality southeast of the Brown Bank, by the crew of the vessel UK33 in March 2000. The specimen is kept in the collection of the Natural History Museum Rotterdam, catalogued as NMR999100001695 (preliminary catalogue number 02-011, as mentioned by Reumer et al. 2003). It is a right mandible (**fig. 13**) with the dental elements p3 and p4. Overall morphology of the dentary reflects known characteristics of *Homotherium latidens*. The specimen, yellow-brownish in colour, is not mineralised at all and lacks all preservational characteristics that are typical of all the other *Homotherium* remains (Early to early Middle Pleistocene) from the North Sea. The area of recovery, southeast of the Brown Bank, is known to produce faunal remains of Late Pleistocene and Early Holocene origin exclusively. In fact, the preservation of the specimen matches that of Late Pleistocene faunal remains recovered from the same area. This strongly suggests a Late Pleistocene date for the specimen, which would make it a unique find, given that it was generally accepted at the time of recovery that *Homotherium* went extinct on the Eurasian continent around 500000 BP (Turner/Antón 1997).

To investigate the age in more detail, the mandible was sampled for ^{14}C dating. Care was taken to obtain non-contaminated samples, as described in detail by Reumer et al. (2003). The results of the radiocarbon analysis (carried out at the R. J. van de Graaff Laboratorium of Utrecht University/NL) were remarkable. The six performed ^{14}C dates, published by Reumer et al. (2003), are reproduced here in **table 1**. These results, *c.* 28000 BP, confirmed our earlier interpretation on the basis of preservation of the specimen and demonstrated definitively the Late Pleistocene survival of *Homotherium* in northwest Europe. Furthermore, the specimen recently yielded aDNA (Paijmans et al. 2015). This can be interpreted as independent evidence for its relatively young age, as aDNA from older material (> 50 kyr) has not yet been obtained from North Sea material.

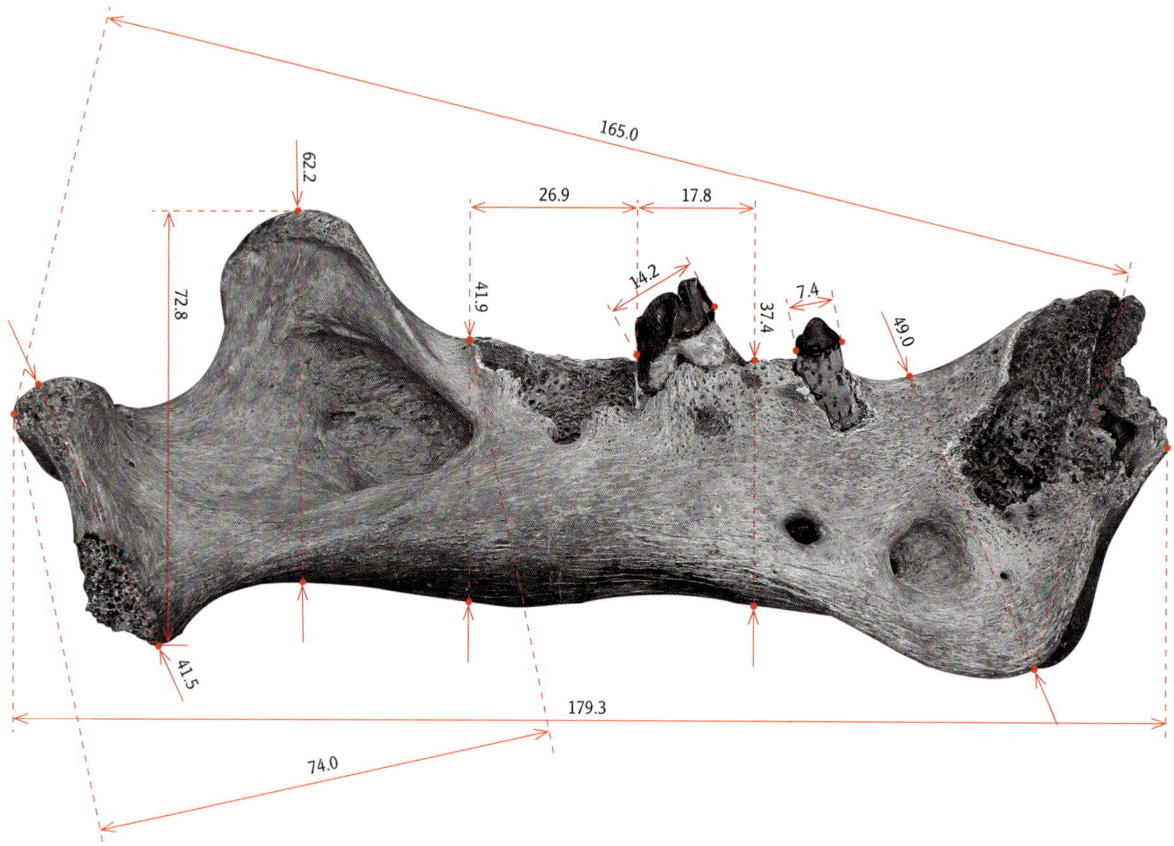

Fig. 13 Buccal side of the Late Pleistocene North Sea *Homotherium latidens* dentary. Measurements in mm. – (Photo H. Wildschut).

UtC number	sample	analysed fraction	mass [mg]	D^{13}C [p mil]	^{14}C age [yr BP]
10456	tooth	collagen1st preparation	2.07	-18.1	31300 ± 400
10999	tooth	collagen 1st preparation	2.33	-17.6	31300 ± 400
10908	mandible	collagen 1st preparation	2.17	-18.9	26900 ± 400
11064	mandible	collagen 1st preparation	1.42	-15.3	26700 ± 240
11000	tooth	collagen 2nd preparation	0.60	-21.2	28100 ± 220
11065	mandible	collagen 2nd preparation	1.01	-17.7	27650 ± 280

Tab. 1 Results of the six performed ^{14}C dates from the North Sea *Homotherium latidens* dentary as published by Reumer et al. (2003). The ^{14}C ages from the second preparation are considered the most reliable.

Inspired by these spectacular ^{14}C results documenting the Late Pleistocene survival of *Homotherium latidens* in northwest Europe, Remie Bakker, a professional sculptor in Rotterdam, decided to produce a life-size reconstruction of this extinct animal (**fig. 14**). The idea was to produce a strolling animal with a powerful head and the typical shape of the lower jaw (**fig. 15**). It was decided to construct the head with its mouth slightly opened, exposing the long and serrated upper canines. The model shows the strongly developed long forelegs and the descending back line, based on the fossil record. Available examples of skeletons of *Homotherium* were used to determine the animal's dimensions. The life-size model with a shoulder height of 1.07 m and a 1.88-m length from the tip of the snout to the tail, is cloaked with artificial fur of different lengths. The main colour chosen was dark grey with an occasional dark stripe, similar to the coat of snow leopards, wolves and lynxes. The idea behind fur colour is that the scimitar cat was an element of

Fig. 14 Life-size model of *Homotherium latidens* by the Dutch sculptor Remie Bakker. – (Photo R. Bleuanus).

Fig. 15 Head of the model. – (Photo R. Bleuanus).

Fig. 16 *Homotherium latidens* resting in the tall grasses of the Late Pleistocene mammoth steppe, now the bottom of the North Sea. – (Painting R. Bakker).

the megafauna of the Late Pleistocene, an era dominated by woolly mammoths and woolly rhinoceroses which roamed the area of what is now the Southern Bight of the North Sea between the British Isles and the Netherlands, at around 28000 BP. It shows the adaptation of the animal to a cold and dry environment of the mammoth steppe, predominantly with tall grasses (**fig. 16**).

CONCLUSIONS

The Southern Bight of the North Sea between the British Isles and the European mainland is extremely rich in faunal remains of Early Pleistocene to Early Holocene age, indicating that Great Britain was once connected to the European mainland. Most of the skeletal elements are very well preserved. They are all *ex situ* finds and either a bycatch of the fishing industry, obtained through expeditions fishing specifically for fossils, or collected during excursions on the artificial beaches alongside the Dutch coastline such as the beaches of Rockanje, Maasvlakte 1 and 2, Hoek van Holland and De Zandmotor. Based on their morphology, these skeletal elements can be identified to species. The species, where well known, can be placed within a certain time span of the Pleistocene or the Early Holocene. Factual data, such as the area where the fossils are retrieved and the geological conditions in combination with the state of mineralisation, can be used confidently for a rough dating of the material. The four specimens from the Maasvlakte 2 artificial beach attributed to the Early Pleistocene *Homotherium latidens* presented here for the first time are good examples of this.

Most of the North Sea remains of the two species of sabre-toothed cats, *Homotherium crenatidens* and *Homotherium latidens*, can be placed in the Early Pleistocene to earliest Middle Pleistocene. The North Sea mandible of *Homotherium latidens* can be easily differentiated from the Early to early Middle Pleistocene North Sea remains: it is not at all mineralised, and the ^{14}C analysis determined its age to be *c.* 28000 BP. So far, that mandible in the collections of the Natural History Museum Rotterdam is the only specimen documenting the Late Pleistocene survival of *Homotherium latidens* in northwest Europe up to 28000 BP. To date, it is the youngest *Homotherium* from the Eurasian continent.

ACKNOWLEDGEMENTS

We wish to thank Marc Simmelink, Walter Langendoen and René Lamers for putting the Maasvlakte 2 specimens of *Homotherium* at our disposal for comparative studies. Wilrie van Logchem provided casts of crania of *Homotherium* and *Xenosmilus* for comparative investigations and contributed to fruitful discussions. René Bleuanus, Frits van der Vossen, Hans Wildschut and Marc Simmelink are thanked for providing images used in this paper. We also wish to express our gratitude to Klaas Post for critically reading a first draft of the manuscript and providing information on the North Sea fossils which have been trawled by the fishing industry over the years. John Jagt and Jeanne Timmons improved the English of this paper. Last but not least, D. Mol wishes to express his gratitude to the organisers for the invitation to participate in the workshop »The *Homotherium* finds from Schoeningen 13II-4 and the big cats of the Ice Age« (June 2015).

REFERENCES

Alter et al. 2015: S. E. Alter / M. Meyer / K. Post / P. Czechowski / P. Gravlund / C. Gaines / H. C. Rosenbaum / K. Kaschner / S. T. Turvey / J. van der Plicht / B. Shapiro / M. Hofreiter, Climate impacts on transocean dispersal and habitat in gray whales from the Pleistocene to 2100. Molecular Ecology 24, 2015, 1510-1522.

Ballesio 1963: R. Ballesio, Monographie d'un *Machairodus* du gisement Villafranchien de Senèze: *Homotherium crenatidens* Fabrini. Travaux du Laboratoire de Géologie de la Faculté des Sciences de Lyon, N. S. 9, 1963, 9-129.

Berendsen 2011: H. J. A. Berendsen, De vorming van het land. Inleiding in de geologie en de geomorfologie (Assen 2011).

Busschers/van Heteren/Westerhoff 2012: F. Busschers / S. van Heteren / W. Westerhoff, Het stratigrafische raamwerk voor de geologische opbouw van het zandwingebied Maasvlakte 2. Rapport 1 van het geo-archeologisch en paleontologisch onderzoek zandwingebied en buitencontour Maasvlakte 2. TNO 2012 R11131 (2012).

Hemmer 2001: H. Hemmer, Die Feliden aus dem Epivillafranchium von Untermaßfeld. In: R.-D. Kahlke (ed.), Das Pleistozän von Untermassfeld bei Meiningen (Thüringen), Teil 3. Monographien des RGZM 40, 3 (Mainz 2001) 699-782.

2004: H. Hemmer, Notes on the ecological role of European cats (Mammalia: Felidae) of the last two million years. In: E. Baquedano Pérez / S. Rubio Jara (eds), Miscelánea en homenaje a Emiliano Aguirre. 2: Paleontología. Zona Arqueológica 4 (Madrid 2004) 214-232.

van Hooijdonk 1999: K. van Hooijdonk, De vondst van de maand. De vondst van een calcaneum of hielbeen van een Pleistocene sabeltandtijger. Cranium 16/2, 1999, 102-104.

Hooijer 1962: D. A. Hooijer, The sabre-toothed cat *Homotherium* found in The Netherlands. Lutra 4, 1962, 24-26.

van Kolfschoten/Laban 1995: T. van Kolfschoten / C. Laban, Pleistocene terrestrial mammal faunas from the North Sea. Mededelingen Rijks Geologische Dienst 52, 1995, 135-151.

Kortenbout van der Sluijs 1971a: G. Kortenbout van der Sluijs, Bones of mammals from the Brown Bank area (North Sea). In: L. P. Louwe Kooijmans, Mesolithic bone and antler implements from the North Sea and from the Netherlands. Berichten van de Rijksdienst voor Oudheidkundig Bodemonderzoek 20-21, 1970-1971 (1971), 69-70.

1971b: G. Kortenbout van der Sluijs, Fossil bones from the North Sea. Netherlands Journal of Zoology 21/2, 1971, 214.

1983: G. Kortenbout van der Sluijs, De resten van zoogdieren uit de Noordzee. Grondboor & Hamer 37, 1983, 4-7.

1985: G. Kortenbout van der Sluijs, Botten uit de Oosterschelde. Cranium 2/1, 1985, 9-10.

Laban/Rijsdijk 2002: C. Laban / K. F. Rijsdijk, De Rijn-Maasdelta's in de Noordzee. Grondboor & Hamer 56, 2002, 60-65.

van Logchem/Mol 2008: W. van Logchem / D. Mol, De vroeg-pleistocene sabeltandkat, *Homotherium crenatidens* (WEITHOFER, 1889), voor het eerst opgevist van de bodem van de Noordzee. Cranium 25/2, 2008, 3-16.

Mol/Langeveld 2015a: D. Mol / B. Langeveld, New fossil discoveries from a dredge-constructed beach: Part 1. Deposits 41, 2015, 26-35.

2015b: D. Mol / B. Langeveld, New fossil discoveries from a dredge-constructed beach: Part 2. Deposits 42, 2015, 26-31.

2016: D. Mol / B. Langeveld, Safari mammoetsteppe. Historyland (Hellevoetsluis 2016).

Mol/Post 2010: D. Mol / K. Post, Gericht korren op de Noordzee voor de zoogdierpaleontologie: een historisch overzicht van de uitgevoerde expedities. Cranium 27/2, 2010, 14-28.

Mol/Reumer 2010: D. Mol / J. W. F. Reumer, The North Sea – Maasvlakte 2 project. In: Volume of Abstracts of the Vth International Conference on Mammoths and their Relatives, Le Puy-en-Velay, August 30 - September 5, 2010. Quaternaire, Hors Série 3, 2010, 196.

Mol et al. 2003: D. Mol / K. Post / J. W. F. Reumer / J. de Vos / C. Laban, Het Gat: Preliminary note on a Bavelian fauna from the North Sea with possible two mammoth species. In: J. W. F. Reumer / J. de Vos / D. Mol (eds), Advances in Mammoth Research. Proceedings of the Second International Mammoth Conference, Rotterdam, 16-20 July 1999. Deinsea 9 (Rotterdam 2003) 253-266.

2006: D. Mol / K. Post / J. W. F. Reumer / H. van der Plicht / J. de Vos / B. van Geel / J. P. Pals / J. Glimmerveen, The Eurogeul – first report of paleontological, palynological and archeological investigations of this part of the North Sea. Quaternary International 142-143, 2006, 178-185.

2008a: D. Mol / J. de Vos / R. Bakker / B. van Geel / J. Glimmerveen / H. van der Plicht / K. Post, Kleine encyclopedie van het leven in het Pleistoceen. Mammoeten, neushoorns en andere dieren van de Noordzeebodem (Diemen 2008).

2008b: D. Mol / W. van Logchem / K. van Hooijdonk / R. Bakker, The saber-toothed cat of the North Sea (Norg 2008).

2013: D. Mol / H. van der Plicht / J. de Vos / B. Gravendeel / W. Langendoen / W. van den Broek / B. Langeveld / F. Dieleman / J. Reumer, The Eurogeul area – new paleontological data of this part of the North Sea. In: International Conference »World of Gravettian Hunters«, Kraków, Poland, 25th-28th June, 2013; Abstracts & Guide Book (Kraków 2013) 58-63.

2015: D. Mol / B. Langeveld / A. Janse / W. Langendoen / J. Smolarz, Determinatiedag fossiele strandvondsten van Maasvlakte 2 in Futureland: een verslag. Cranium 32/1, 2015, 49-58.

Paijmans et al. 2015: J. L. A. Paijmans / J. Reumer / R. Barnett / N. Rohland / J. A. Leonard / A. Barlow / M. Hofreiter, Late Pleistocene survival of *Homotherium* in North Western Europe. In: Programme & Abstracts 21st International Cave Bear Symposium 2015, 11.

Reumer et al. 2003: J. W. F. Reumer / L. Rook / K. van der Borg / K. Post / D. Mol / J. de Vos, Late Pleistocene Survival of the Saber-toothed Cat *Homotherium* in Northwestern Europe. Journal of Vertebrate Paleontology 23/1, 2003, 260-262.

Scager 2015: D. J. Scager, Kor & Bot: 65 jaar en 2000 fossielen later. Een inventarisatie en interpretatie van de collectie. Cranium 32/1, 2015, 6-18.

Turner/Antón 1997: A. Turner / M. Antón, The Big Cats and their Fossil Relatives (New York 1997).

SUMMARY / ZUSAMMENFASSUNG

On Remains of *Homotherium* from the Bottom of the North Sea between the British Isles and the Netherlands and their Stratigraphical Age

This paper focuses on remains of sabre-toothed cats that have been recovered from sediments of the Southern Bight of the North Sea between the British Isles and the Netherlands. These skeletal elements have been recovered by fishermen over the years during trawling for bottom-dwelling flatfish. Other important sources of such fossils are the constructed beach of Maasvlakte 2 and some other replenished beaches. These beaches are being actively constructed by dredging sediments, inclusive of abundant fossil remains from the North Sea floor and dumping them in order to extend coastal land areas. All these remains are thus retrieved *ex situ*. Based on their morphology and state of preservation, these remains can be confidently dated as Early or early Middle Pleistocene, Late Pleistocene and Early Holocene. We discuss the faunal compositions and their age, inclusive of a mandible of *Homotherium latidens* (Owen, 1846) that was collected on March 16, 2000, by the fishing vessel UK33, southeast of the Brown Bank in the North Sea halfway between IJmuiden (the Netherlands) and Lowestoft (East Anglia) and described by Reumer et al. (2003). This specimen was radiocarbon dated by Utrecht University at *c.* 28 000 BP, thus furnishing proof of the survival of *Homotherium latidens* into the Late Pleistocene in northwest Europe. Furthermore, we present four new records of Early Pleistocene *Homotherium latidens*, all recovered recently (2015-2016) from the beach of Maasvlakte 2.

Über Reste von *Homotherium* vom Grund der Nordsee zwischen den Britischen Inseln und den Niederlanden und ihr stratigraphisches Alter

In diesem Beitrag geht es um Überreste von Säbelzahnkatzen, die aus den Sedimenten der südlichen Bucht der Nordsee zwischen den Britischen Inseln und den Niederlanden geborgen wurden. Diese Skelettelemente wurden im Laufe der Jahre von Fischern bei der Schleppnetzfischerei auf bodenbewohnende Plattfische geborgen. Weitere wichtige Quellen für solche Fossilien sind der künstlich aufgeschüttete Strand von Maasvlakte 2 und einige andere aufgespülte Strände. Diese Strände werden aktiv aufgebaut, indem Sedimente, die reichlich fossile Überreste enthalten, vom Nordseeboden gebaggert und verklappt werden, um die Küstengebiete zu erweitern. Alle diese Überreste werden also *ex situ* geborgen. Aufgrund ihrer Morphologie und ihres Erhaltungszustands können diese Überreste mit Sicherheit in das frühe oder frühe Mittelpleistozän, das Spätpleistozän und das frühe Holozän datiert werden. Wir diskutieren die Zusammensetzung der Fauna und ihr Alter, einschließlich eines Unterkiefers von *Homotherium latidens* (Owen, 1846), der am 16. März 2000 vom Fischereischiff UK33 südöstlich der Brown Bank in der Nordsee auf halbem Weg zwischen IJmuiden (Niederlande) und Lowestoft (East Anglia) gesammelt und von Reumer et al. (2003) beschrieben wurde. Dieses Exemplar wurde von der Universität Utrecht auf ca. 28 000 BP ^{14}C-datiert und liefert damit den Beweis für das Überleben von *Homotherium latidens* bis ins Spätpleistozän in Nordwesteuropa. Außerdem stellen wir vier neue Nachweise von *Homotherium latidens* aus dem frühen Pleistozän vor, die alle kürzlich (2015-2016) am Strand von Maasvlakte 2 gefunden wurden.

ROSS BARNETT

A VERY BRITISH SABRETOOTH: *HOMOTHERIUM* IN THE UK

The sabretooth cats (Felidae, subfamily Machairodontinae) are the extinct cousins of the modern conical-toothed cats (Felidae, subfamily Felinae; Barnett et al. 2005). The stocky New-world *Smilodon fatalis* is probably the most familiar species, known from thousands of specimens recovered from the Pleistocene tarpits of Rancho la Brea in downtown Los Angeles (Merriam/Stock 1932). Sympatric with *Smilodon*, although much rarer in the fossil record is the »scimitar-tooth« cat *Homotherium serum*, typical of the genus, with a more athletic build, clearly adapted for high-speed pursuit of prey, rather than ambush, and with canines that are broad and coarsely serrated (Antón/Galobart 1999; Rawn-Schatzinger 1992; Turner/Antón 1997). The tribe Homotheriini represent a very successful radiation of the sabretooth lineage, with species identified in Africa (Ewer 1954), Europe (Arribas/Garrido 2008; Reumer et al. 2003), Asia (Teilhard de Chardin 1939; von Koenigswald 1934), North America (Jefferson/Tejada-Flores 1993) and, most recently, in South America (Mones/Rinderknecht 2004; Rincón/Prevosti/Parra 2011).

Within Europe, *Homotherium* has generally been subdivided into two chronological grades: *H. crenatidens* from the Early Pleistocene and *H. latidens* from the Middle and Late Pleistocene. Separation between the two has mainly been made on the basis of size and canine shape (Antón/Galobart/Turner 2005; Turner/Antón 1997).

The distinctively serrated triangular canines that we now recognize as belonging to *Homotherium* were first found at the famous Italian site of Val d'Arno (prov. Pisa/I), where they were initially ascribed to a new species of bear: *Ursus cultridens*. It was using this description that the Rev. John MacEnery identified and figured a number of canines from the site of Kent's Cavern (Devon). These British canines then formed the basis for Richard Owen's description of the first sabretooth cat, *Machairodus latidens* (Owen 1846), now recognized as *Homotherium latidens*. Since then, *Homotherium* has been found at the site of Robin Hood Cave, Creswell Crags (Derbyshire); Victory Quarry, Dove Holes, Buxton (Derbyshire); Westbury-sub-Mendip (Somerset); and from the erosional deposits of East Anglia (**fig. 1**). Particular confusion has arisen over the Kent's Cavern material due to MacEnery's untimely death and the posthumous dispersal of both his writings and his precious fossils. As important as they are, the lack of a cohesive understanding of what material was found has so far hindered a complete understanding of both the site and the species.

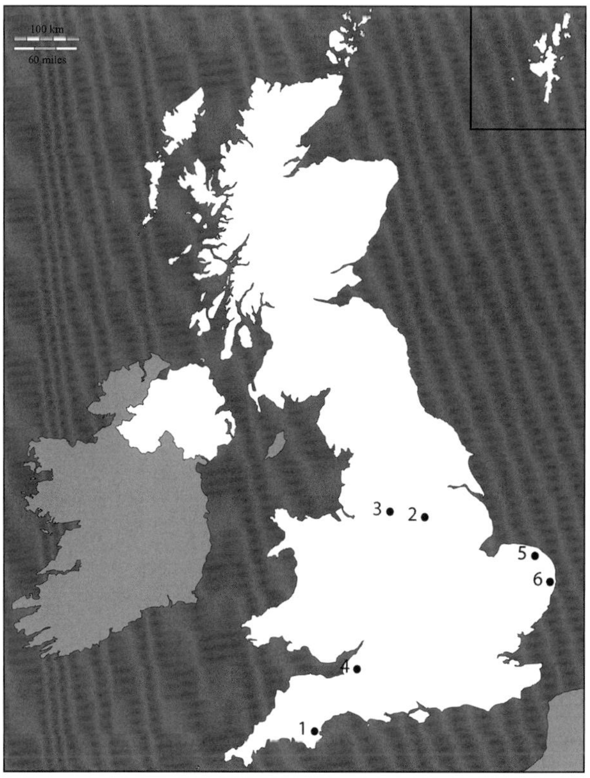

Fig. 1 UK map with location of sites mentioned in the text: **1** Kent's Cavern, Devon. – **2** Creswell Crags, Derbyshire. – **3** Dove Holes, Derbyshire. – **4** Westbury-sub-Mendip, Somerset. – **5** Cromer, Sidestrand, West Runton and Bacton, Norfolk. – **6** Kessingland, Pakefield, and Covehithe, Suffolk. – (Map modified from www.d-maps.com).

In: Nicholas J. Conard · Henning Hassmann · Kurt Felix Hillgruber · Jordi Serangeli · Thomas Terberger (eds), The *Homotherium* Finds from Schöningen 13II-4. Man and Big Cats of the Ice Age. Forschungen zur Urgeschichte aus dem Tagebau von Schöningen 4 (Mainz 2022). DOI: https://doi.org/10.11588/propylaeum.1006.c13521

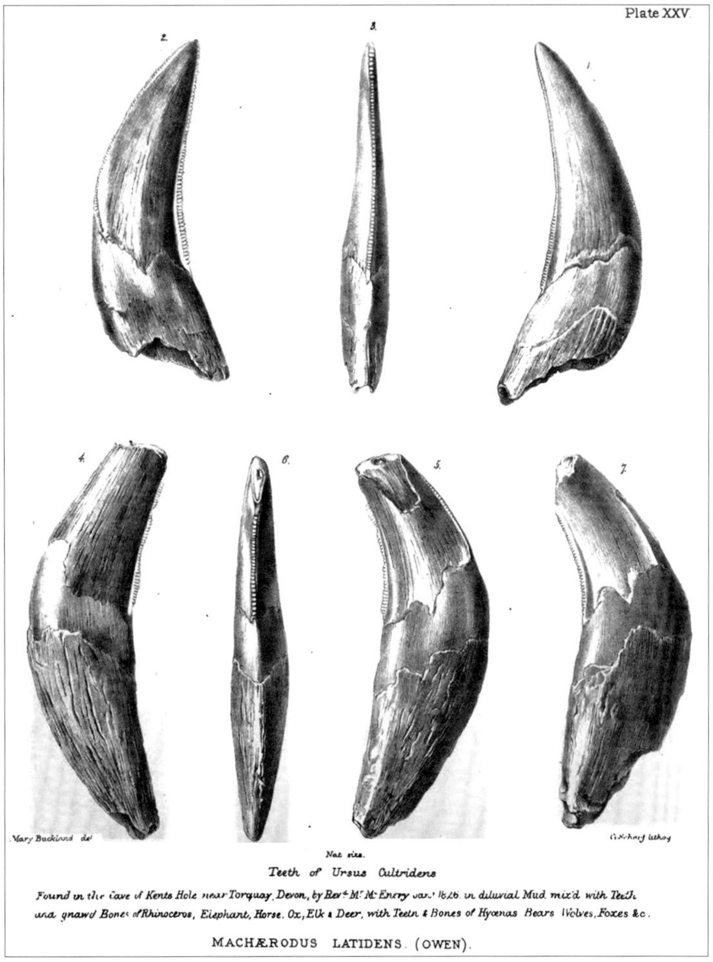

Fig. 2 Plate F of MacEnery, delineated by Mary Buckland and lithographed by George Scharf. – (After Dawkins/Sanford/Reynolds 1878, pl. XXV).

CORPUS OF FINDS IN THE UK

Kent's Cavern, Devonshire

Canines

The Rev. John MacEnery (1796-1841; Clark 1925a) was the private chaplain to the Cary family of Tor Abbey (Clark 1925b). He took a keen amateur interest in the caves of the area and was the first person to attempt any serious excavation of the site now known as Kent's Cavern. In January 1826, while digging in the area known as »the wolf's den« he encountered five canines of a type unknown to him (Pengelly 1868). These were passed to Prof. William Buckland of the University of Oxford, who was also unable to identify them. They were finally recognised by Baron Georges Cuvier as being identical with his »*Ursus cultridens*« from Val d'Arno (Pengelly 1868). Very early in 1826, Mary Buckland (William's wife) drew a number of the unique canines in preparation for a high quality lithograph. The final result was the beautiful Plate F (**fig. 2**), which MacEnery intended to publish in full with his excavation notes on Kent's Cavern in a book to be called »Cavern Researches« (Pengelly 1868). Unfortunately, MacEnery died before completing the work and his collection and manuscripts were scattered by auction in 1842 (Pengelly 1868). Several attempts at publishing his findings as a single coherent work were made posthumously (MacEnery 1859; Pengelly 1868) but it was not until 1878 that Plate F become widely circulated when it was printed as Plate XXV of Dawkins and Sanford's (1878) classic »British Pleistocene Mammalia«.

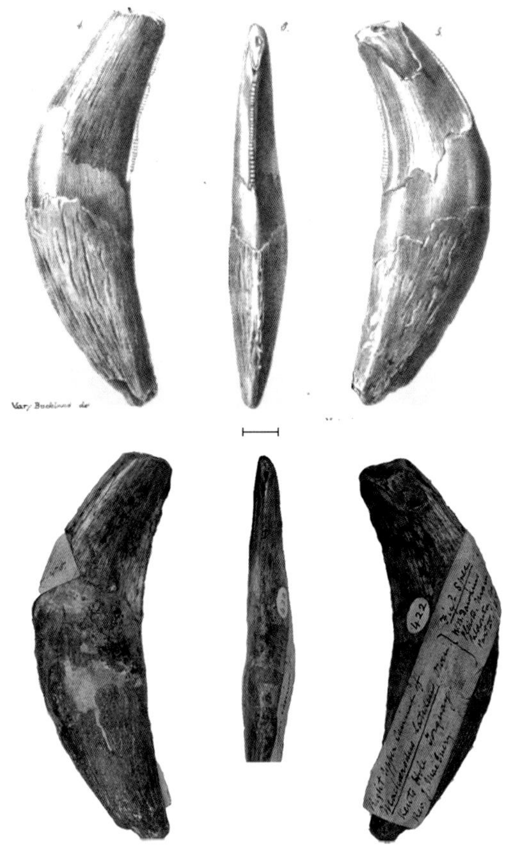

Fig. 3 Comparison of Institute of Geological Science Museum GS422 canine with MacEnery's Plate F, figs 4-6. – (Photo British Museum). – Scale Bar 10 mm.

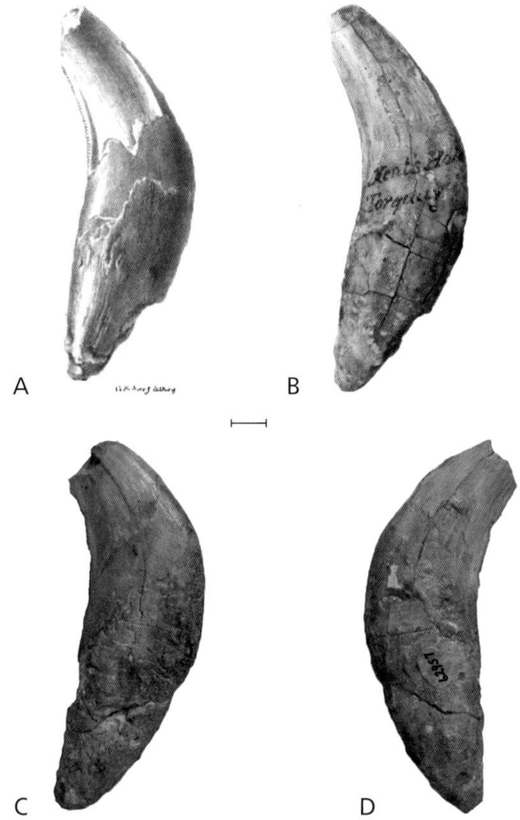

Fig. 4 Comparison of Oxford University Museum of Natural History Q955 canine and Institute of Geological Science Museum IGS62957 canine with MacEnery's Plate F fig. 7 – **A** Plate F, fig. 7. – **B** Oxford University Museum of Natural History Q955 canine. – **C-D** Institute of Geological Science Museum IGS62957 canine: **C** lingual view; **D** labial view. – Scale Bar 10 mm.

Since no legend was found with the plate, there has been confusion over how many canines are represented, whether differing views of the teeth represent the same specimen or not, and where the fossils have finally ended up. MacEnery's notes (P197, Fasciculus D Pengelly 1868) only refer to »P.F Fig 1.2.3-exhibiting different views of the most perfect tooth«, but the figures 4, 5, 6, and 7 could potentially have represented the remaining four canines from various angles. Over the years, several authors have offered their interpretation of what was represented in Plate F. Dawkins and Sanford (1878) state »[...] the original of pl. F' Figs. 4, 5, found its way into the museum of the geological society; the fourth, figured, pl. F' figs. 1, 2, 3, is in the British museum; and the fifth, (pl. F' fig. 7) is in the collection of Sir Walter Trevelyan, Bart.«

This information is elaborated upon by Kennard (1945), who correctly states that figures 4, 5 and 6 are of the same tooth (**fig. 3**) but that figure 7 is the canine now in the Geological Survey Collections after donation by Sir W. C. Trevelyan in 1871. Crucially however, it appears that both Dawkins and Kennard were misled and that figure 7 is actually the canine in the Oxford University Museum of Natural History (**fig. 4**). Presumably, neither had seen the Geological Survey or Oxford specimens for comparison to Plate F. Now, for the first time since MacEnery died in 1841 we have the correct identification of all three canines figured in Plate F, and where they are currently curated (**tab. 1**). The fourth canine, now housed at the Institute of Geological Science, London (IGS62957; **tabs 1. 3; fig. 4, C-D**), has never been figured before.

Kent's Cavern is the type locality for *Homotherium latidens*, with a holotype described by Owen in 1846. Although he does not explicitly identify the canine in his diagram and description, it is clearly canine 443/103

no.	element	museum	accession	figured	notes	purchase history	references
1	left upper canine	British Museum of Natural History	14954	Dawkins 1878, pl. 25 figs 1-3 Dawkins 1874a, fig.	**Kennard A** (1945) casts in Paris, Oxford (Q7261)	MacEnery Auction 1842 → Lovell Phillips (Falconer 1868) → König (Kennard 1945)	Falconer 1868 Dawkins 1874b; 1878 MacEnery 1859 Pengelly 1869; 1871 Kennard 1945 Lyell 1873
2	right upper canine	Geological Society (now IGS, London)	GS422 (23,413)	Dawkins 1878, pl. 25 figs 4-6 (Kennard 1945)	**Kennard B** (1945) casts in Oxford (Q7262) and NHM (1868) and Paris (1945)	Mrs Cazalet donated Feb. 16th, 1826 (Falconer 1868), or Feb. 17th, 1826 (Kennard 1945)	Dawkins 1878 Kennard 1945 Falconer 1868
3	right upper canine	Museum of Practical Geology (now IGS, London) (1983) Geological Survey	62957	wrongly identified as pl. 25 fig. 7 by Dawkins/Sanford 1878 and Kennard 1945	**Kennard C** (1945)	Mrs Cazalet (in 1826) → Sir W. C. Trevelyan donated May 15th 1871 (Kennard 1945)	Dawkins 1878 Cleevely 1983 Kennard 1945
4	right upper canine	Museum of the Royal College of Surgeons	443 or 103 canine destroyed in 1941 (Cleevely 1983)	Owen 1846, P.180 fig. 69 (Lydekker 1885; Owen 1846)	**Kennard D** (1945) cast: NHM 46842 (Lydekker 1885) 46842a (Lydekker 1885) cast in Albert Museum, Exeter? (Cleevely 1983)	MacEnery Auction 1842 → Dr. Battersby → Lord Enniskillen (Falconer 1868)	Dawkins 1878 Cleevely 1983 Kennard 1945 Owen 1846 Lydekker 1885 Falconer 1868 Anon. 1865
5	upper canine	Oxford Museum	Q955	Dawkins 1878, pl. 25 fig. 7	**Kennard E** (1945) cast in York	MacEnery Auction 1842 → Lovell Phillips→ Buckland (Falconer 1868)	Dawkins 1878 Kennard 1945 Falconer 1868
6	3rd left upper incisor	Albert Museum, Exeter	Fos25 now lost? (Dawkins 1869)	Owen 1846, P.182 fig. 70 Dawkins 1878, P.188 Dawkins 1874a, figs 103-105 Gervais 1848-1852, P.126 Gervais 1867-1869, P.78 fig. 4	**Kennard 1, 2, 3, 4** cast: NHM 46767 cast in Paris (1945)	MacEnery Auction 1842 → W. C. Radley → F. W. L. Ross (in 1853) (Kennard 1945)	Kennard 1945 Dawkins 1874a; 1878 Owen 1846 Pengelly 1875 Gervais 1848-1852
7	3rd right upper incisor	British Museum of Natural History (NHM)	M582	Pengelly 1897, P.227 Lyell 1873, P.105 fig. 10	**Kennard 5** July 29, 1872, found	Pengelly → Lord Haldon → donated in 1883	Pengelly 1883 Lydekker 1885 Pengelly 1897 Kennard 1945

Tab. 1 The Kent's Cavern material.

from the Royal College of Surgeons. Sadly, this identification had to be made through consulting casts 46842 and 46842a in the British Museum of Natural History as the original was destroyed during bombing of the Royal College in 1941. This is the fifth and final canine of the five discovered by MacEnery. Owen (P181, 1846) mentions »Three of these canine teeth [...] were discovered by the Rev. Mr. MacEnery in Kent's hole, Torquay«, although we know that five canines were found during MacEnery's excavations, any of these paratypes could be elevated to the rank of neotype.
Recent research on the canines has suggested that the teeth are not native to the Devon area but were instead transported there by Palaeolithic people for reasons unknown (McFarlane/Lundberg 2013). If the scimitar-cat held cultural significance for Palaeolithic groups then it is likely that the concentration of canines and incisors from this species in Kent's Cavern are from somewhere else. The source site is currently unknown and whether the teeth were taken from a recently dead *Homotherium* or from subfossil material remains a matter of conjecture. With the discovery that the canines have been imported (possibly as fossils from the continent), a crucial piece of evidence for Late Pleistocene survival of *H. latidens* in North-Western Europe is removed. It is now only the equally contentious remains from Creswell Crags that provide support for Late Pleistocene *Homotherium* in Britain.

Incisors

During Pengelly's systematic excavations of Kent's Cavern in the 1860s and 1870s, it was hoped that further remains of *Homotherium* would be discovered, and the earlier findings of MacEnery would be confirmed. Despite eight seasons of digging from 1864 to 1872, no trace of the elusive sabretooth was found. Finally, on 29 July, 1872 »a well-marked incisor of *Machairodus* [Homotherium] *latidens*« was found, »with a left ramus of lower jaw of bear, containing one molar, in the first of uppermost foot-level of cave-earth, having over it the granular stalagmitic floor 2.5 feet deep« (**fig. 5**; Pengelly 1873c). This is the only *Homotherium* tooth from Kent's Cavern for which there is a comprehensive account of the discovery and curation. Pengelly bequeathed the incisor to Lord Haldon (owner of Kent's Cavern) who donated the tooth to the British Museum (Natural History) in 1883 (Lydekker 1885).
In MacEnery's notes (MacEnery 1859) he mentions »I have lately discovered in the same bed a small tooth about an inch long – the internal face of the enamel is fringed with a serrated border-this tooth is distinguished farther by two tubercles or protuberances at the base of the enamel from which the serration springs and describes a pointed arch on the internal surface vid – fig. 8 9«. This description of what is clearly an incisor of *Homotherium*, has been the starting point for a long and confused discussion of how many incisors were actually found by MacEnery in Kent's Cavern. As this is the only mention made of an incisor, and given the number of references to the canines throughout his manuscripts it is logical to think that only one incisor was found by MacEnery. However, several authors have contested this straightforward assessment, with up to four separate incisors claimed to have come from MacEnery's diggings in Kent's Cavern (Kennard 1945). The evidence for multiple incisors found by MacEnery hinges on the interpretation of two images and how they relate to a *Homotherium* incisor in the Royal Albert Memorial Museum (RAMM), Exeter (Kennard 1945; Pengelly 1875).
The RAMM incisor has had a very interesting history (see Pengelly 1875) and is certainly the tooth found by MacEnery, although it has been roughly treated since, and has lost some of the enamel and root it originally retained on discovery. The characteristic serration of the enamel is still obvious and diagnostic.
The first image to introduce confusion is that on P188 of the »British Pleistocene Mammalia« (Dawkins/Sanford/Reynolds 1878). Dawkins/Sanford reproduce a woodcut (based on a photo of a drawing in Indian ink)

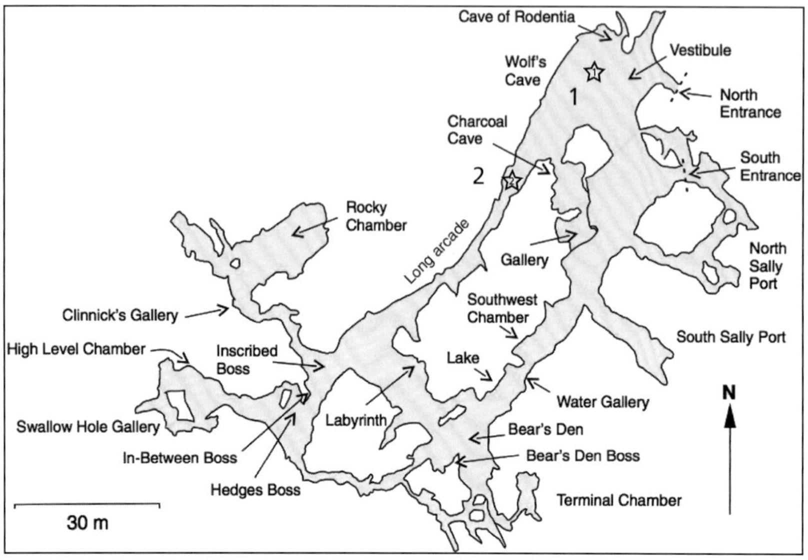

Fig. 5 Plan of Kent's Cavern from Lundberg (Lundberg/McFarlane 2007, 200) with approximate sites (star) of *Homotherium* remains identified. – **1** MacEnery's five canines and one incisor from the »wolf's den«. – **2** Pengelly's incisor from the long arcade.

of an image found in the collection of MacEnery after his death. The original drawing (currently untraced) contained five images; three of *Homotherium* incisors and two of the upper jaw of a horse, apparently drawn by G. Scharf in 1837 (Pengelly 1873b). **Figure 6, A** has mostly been interpreted as representing two incisors (Kennard 1945; Pengelly 1873a), with »fig. 1« and »fig. 2« representing left and right sides of one specimen and »fig. 3« a separate specimen. Kennard identifies these hypothetical incisors as numbers 3 and 4 in his system (Kennard 1945). However, it is my opinion that despite the rather unusual dimensions portrayed by the separate figures, only one incisor is represented in three views, in an analogous manner to the images of Pengelly's incisor in Lyell (1873) (see **fig. 6, E** for comparison). The proportions appear to correspond in all views (e. g. between cusps and apex in the crown, and between crown and root). The striations present on the root, presumably due to hyaena (*Crocuta crocuta spelaea*) gnawing, and therefore uniquely characteristic, appear to match. Interestingly, Pengelly, who was the first to have access to the Indian ink drawing in his capacity as president of the Torquay Natural History Society, initially states »amongst the plates [...] there is one containing five figures, of which three are without doubt different views of the tooth described by the author [MacEnery]« (P198, Pengelly 1868). In subsequent work (Pengelly 1872; 1873a), he appears to defer to the opinion of Dawkins, who discusses the image as containing two incisors (Dawkins 1874a; 1874b; Dawkins/Sanford/Reynolds 1878; Pengelly 1872).

The second and only other image of a claimed *Homotherium* incisor from Kent's Hole is that of Owen's »A history of British fossil mammals and birds« (Owen 1846). Pengelly identifies this figure as identical to the incisor found by MacEnery (Pengelly 1873b i. e. the RAMM incisor), which was in private collection at the time of his writing. However, the RAMM incisor is a left I3, whereas Owen's figure is apparently of a right I3. This difference is likely due to accidental mirroring during the process of printing as suggested by Burmeister and Pengelly (Pengelly 1875), and Owen himself believed that his image and the RAMM incisor were the same (P260, Pengelly 1875). Additionally, Pengelly notes that whereas the Dawkins woodcut image shows clear signs of hyaena gnawing, both the RAMM incisor and Owen's figure do not have this feature. Pengelly provides a probable explanation for this discrepancy when he states of the RAMM incisor: »there are no traces of teeth marks on it, but it must be observed that it has lost a portion of the surface on one side of the fang [root], obviously since its exhumation. The aspect of the scar suggests that the specimen had been fastened with strong cement to a tablet, and that it had been roughly detached« (P252, Pengelly 1875). As the Owen figure must have been drawn on or after 1844, and MacEnery's specimens,

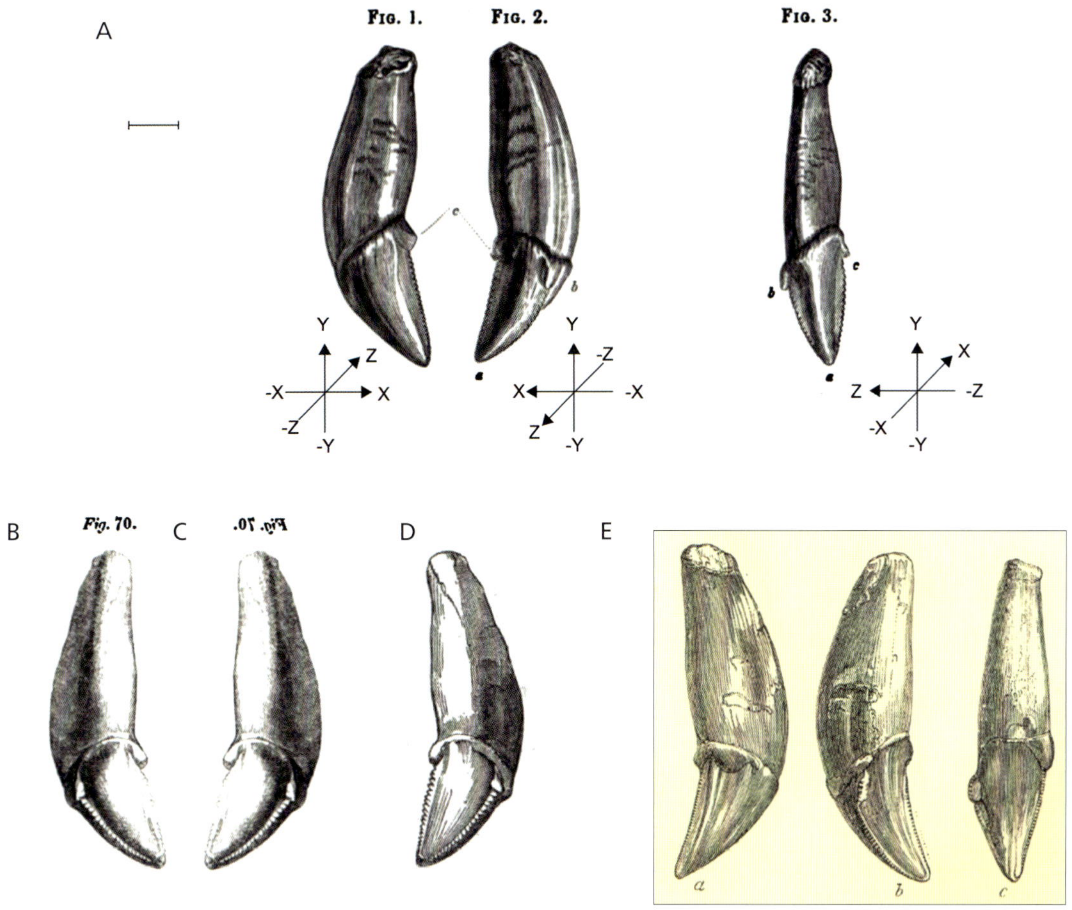

Fig. 6 **A** copy of woodcut figure of incisor in Dawkins and Sanford with apparent orientation axes marked. – **B** incisor from Owen. – **C** Mirrored version of the incisor from Owen. – **D** woodcut of incisor from Gervais, obviously a reflected version of Owen's. – **E** three views of Pengelly's 1872 incisor from Lyell (note similarity in orientations to **A**). – (A after Dawkins/Sanford/Reynolds 1878, fig. 1-3; B-C after Owen 1846, fig. 70; D after Gervais 1848-1852, pl. 32; E after Lyell 1873, fig. 10). – Scale Bar 10mm.

mostly stuck to cards (P491, Pengelly 1869), were dispersed at auction in 1842, the reason for the difference between the images of Dawkins and Owen, and the RAMM incisor become clear.

In summary, a probable timeline of the RAMM incisor is as follows:

1. Found by MacEnery in Kent's Cavern on or before 1837;
2. MacEnery commissions Scharf to produce some preparatory sketches of the incisor from a number of angles with the idea of introducing two views into Plate F as Figures 8 and 9. This is the origin of the Indian ink sketch published by Dawkins and Sanford.
3. MacEnery dies 1841, before publishing his work on Kent's Cavern and his collection is dispersed at auction in 1842.
4. W. C. Radley of Newton Abbot buys the incisor amongst a lot of MacEnery's fossils. After purchase, the specimen is roughly removed from its auction card and loses some portions of root and the tip of the crown.
5. At some point around 1844, the incisor is brought to William Buckland and figured by Richard Owen.
6. F. W. L. Ross of Topsham buys the incisor from W. C. Radley in 1853 and bequeaths it to the RAMM.
7. The incisor is recognised as belonging to *Homotherium,* by W. S. M. D'Urban and described by Pengelly in 1875.

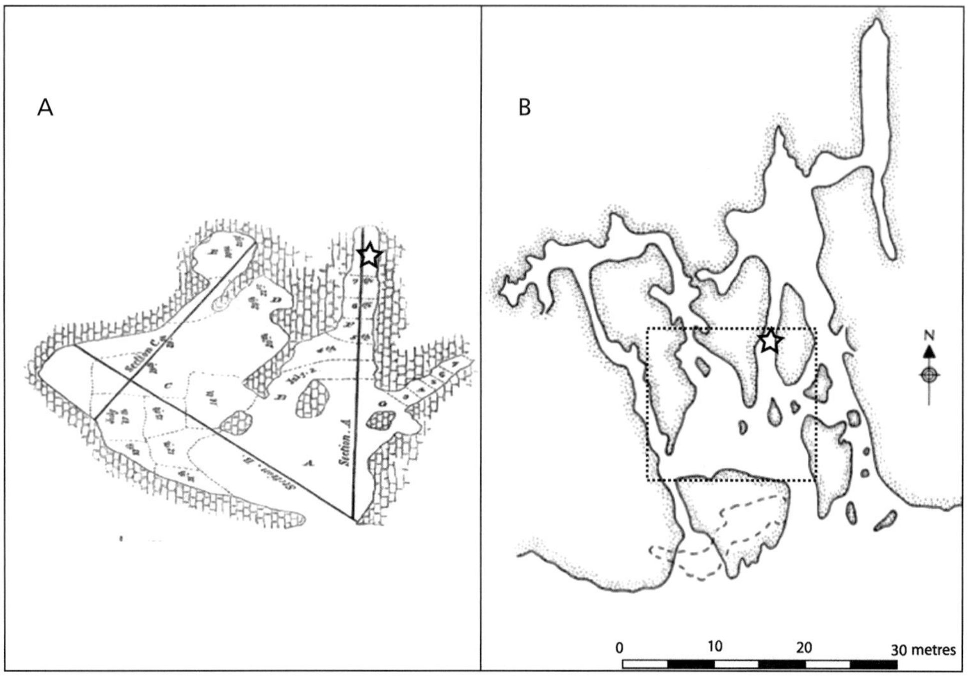

Fig. 7 **A** plan of Robin Hood Cave from Mello with site of *Homotherium* canine identified (star) »at the far end of chamber F«. – **B** plan of modern Robin Hood Cave adapted from Charles/Jacobi with Mello's plan identified by dashed lines. – (A after Mello 1877, fig. 8; B after Charles/Jacobi 1994, fig. 2).

Robin Hood Cave, Creswell Crags, Derbyshire

Second only in controversy to the Kent's Cavern material, the Creswell Crags canine has been the subject of much speculation (Anon. 1879-1880; Charles/Jacobi 1994; Mello 1879-1880; **tab. 2**). Unearthed in the presence of W. Boyd Dawkins, there is a huge literature of claim and counter-claim both for and against the authenticity of the canine (Anon. 1879-1880; Mello 1879-1880). The canine is enigmatic and apparently out-of-place in the Late Pleistocene context of Robin Hood Cave. However, isotopic studies have tentatively confirmed the canine as part of the Creswell assemblage (Oakley 1980) while the recent dating of North Sea material to c. 28000 BP has removed much, but not all, of the objection to *Homotherium* survival into the Late Pleistocene (Reumer et al. 2003).

The canine itself was found in »chamber F« (**fig. 7**) during excavation in 1876 (Mello 1877). Dawkins (1877) states that »it lay about one foot below the stalagmite in the cave-earth; and in association with it were a fine flint flake and remains of bear, woolly rhinoceros, reindeer, horse, and mammoth«.

Victory Quarry, Dove Holes, Buxton, Derbyshire

Discovered in 1901, the infilled cave from Victory Quarry produced a spectacular assemblage of fossil mammals (**tab. 3**). Remains of *Homotherium* include one right radius, one right tibia, three upper canines, two left upper carnassials, and two humeri. As well as *Homotherium*, remains of hyaena, mastodon, mammoth, rhino, horse and deer were also collected and identified by W. B. Dawkins (1903). Originally described as being Pliocene in age by Dawkins (1903), this age has been revised significantly by later authors and is now thought to be Early Pleistocene (Spencer/Melville 1974; Stuart 1974). This material currently represents the earliest appearance of *Homotherium* in the British Isles and therefore should probably be recognised as *H. crenatidens*.

element	museum	accession	figured	notes	references
upper canine	Manchester, currently Creswell Visitor Centre	P.1787	Dawkins 1877, fig. 3	discovered on 3rd July 1876, Chamber F	Dawkins 1877 Jenkinson 1984 Mello 1877 Mello 1880

Tab. 2 The Creswell Crags material.

no.	element	museum	accession	figured	notes	references
1	upper canine	Manchester Museum	L6190e			
2	upper canine	Manchester Museum	L6190d	Dawkins 1903, pl. 9 fig. 1		Dawkins 1903 Spencer 1974 Nudds 1992
3	upper canine	Manchester Museum	L6190a	Dawkins 1903, pl. 8 fig. 1 Spencer 1974, pl. 2 fig. 1		Dawkins 1903 Spencer 1974 Nudds 1992
4	left upper carnassial	Manchester Museum	L6190b	Dawkins 1903, pl. 8 fig. 2 Spencer 1974, pl. 2 fig. 3		Dawkins 1903 Spencer 1974 Nudds 1992
5	left upper carnassial	Manchester Museum	L6190c	Dawkins 1903, pl. 8 fig. 3 Spencer 1974, pl. 2 fig. 4		Dawkins 1903 Spencer 1974 Nudds 1992
6	right radius	Manchester Museum	LL.4126			Dawkins 1903 Spencer 1974 Jackson 1952
7	right tibia	Buxton Museum	no accession number	Dawkins 1903, pl. 11 fig. 1 Spencer 1974, pl. 2 fig. 5	cast in NHM M10171 (Anon. 2020) cast in Manchester Museum L.6192	Dawkins 1903 Spencer 1974 Anon. 2020
8	femur	Manchester Museum	L6197	Dawkins 1903, pl. 11 fig. 4	actually a horse metatarsal (1974)	Dawkins 1903 Spencer 1974
9	distal humerus	Manchester Museum	LL4021			Spencer 1974 Jackson 1952
10	humerus	Manchester Museum	L.6221		gnawed by hyaenas	

Tab. 3 The Victory Quarry material.

Westbury-sub-Mendip, Somerset

A filled cave uncovered during quarrying in 1969 and systematically excavated from 1969-1984 (Bishop 1982; Turner 1999; **tab. 4**). *Homotherium* material includes eight canines, two incisors, two molars, and one premolar. The dating of this site has proved problematic with best estimates placing it as of probable Cromerian age (Middle Pleistocene; Bishop 1982; Stuart 1974; Turner 1999).

East Anglia

The East Anglian *Homotherium* material differs from the previously described sites as it comes from an essentially open-air bed rather than a cave assemblage (**tab. 5**). As such, the remains have been collected piecemeal by amateur and professional excavators over the past 140 years with concomitant difficulty in assigning secure contexts (Lankester 1869; Stuart 1974). Several finds have been mentioned only in passing in the early literature and never figured, leading to the possibility that they were misidentifications or were subsequently lost to private ownership. Significant finds include a right mandible from Kessingland/Pakefield (Suffolk; Backhouse 1886) and a third metatarsal and calcaneum from West Runton (Norfolk; Lewis/Pacher/Turner 2010).

CONCLUSION

Although Kent's Cavern is the type locality for *Homotherium latidens*, there is an under-appreciation of how comparatively rich the UK is in this rare felid. The majority of the material has often been neglected in discussions of Pleistocene mammals, probably due to the muddled discussion of 19th-century finds and confusion over what has actually been found. With the reappraisal of the Kent's Cavern material and the inventory presented herein, future researchers can focus more readily on British *Homotherium*.

Note

This paper is an abbreviated version of Barnett 2014.

ACKNOWLEDGEMENTS

Sincere thanks to Jelle W. F. Reumer and Donald McFarlane for useful comments on the manuscript. – Many thanks to Nigel Monaghan (National Museum of Ireland), Sarah Pearson (Royal College of Surgeons), Ros Westwood (Derbyshire Museums), Claudia Hildebrandt (Royal Albert Museum), David Gelsthorpe (Manchester Museum), Malgosia Nowak-Kemp / Eliza Howlett (Oxford University Museum of Natural History), Paul Shepherd (British Geological Survey) and to Andy Currant, Tony Stuart, Lars Werdelin, Nobuyuki Yamaguchi and Blaire Van Valkenburgh for constructive comments and encouragement.

REFERENCES

Anon. 1865: Anonymous, Synopsis of the contents of the museum of the Royal College of Surgeons, England (London 1865).

Anon. 1879-1880: Anonymous, Creswell Cave. Discovery of the *Machairodus* Tooth. Transactions of the Manchester Geological Society 15/17, 1879-1880, 367.

Anon. 2020: Anonymous. www.nhm.ac.uk/research-curation/collections/departmentalcollections/palaeontology-collections/search/index.php (9.9.2020).

Antón/Galobart 1999: M. Antón / A. Galobart, Neck Function and Predatory Behaviour in the Scimitar Toothed Cat *Homotherium latidens* (Owen). Journal of Vertebrate Paleontology 19/4, 1999, 771-784.

Antón/Galobart/Turner 2005: M. Antón / A. Galobart / A. Turner, Co-existence of Scimitar-toothed Cats, Lions and Hominins in the European Pleistocene. Implications of the Post-cranial Anatomy of *Homotherium latidens* (Owen) for Comparative Palaeoecology. Quaternary Science Reviews 24, 2005, 1287-1301.

Arribas/Garrido 2008: A. Arribas / G. Garrido, Machairodontinae (*Megantereon cultridens roderici* n.ssp. and *Homotherium latidens* (Owen, 1846) at the Fonelas P-1 site (Guadix Basin, Granada). In: A. Arribas (ed.), Vertebrados del Plioceno Superior terminal en el suroeste de Europa: Fonelas P-1 y el Proyecto Fonelas. Cuadernos del Museo Geominero 10 (Madrid 2008) 251-277.

Backhouse 1886: J. Backhouse, On a Mandible of *Machaerodus* from the Forest-bed. Quarterly Journal of the Geological Society 42, 1886, 309-312.

Barnett 2014: R. Barnett, An Inventory of British Remains of *Homotherium* (Mammalia, Carnivora, Felidae), with Special Reference to the Material from Kent's Cavern. Geobios 47/1-2, 2014, 19-29.

Barnett et al. 2005: R. Barnett / I. Barnes / M. J. Philips / L. D. Martin / C. Richard Harington / J. A. Leonard / A. Cooper, Evolution of the Extinct Sabretooths and American Cheetahlike Cat. Current Biology 15/15, 2005, R589-R590.

Bishop 1982: M. J. Bishop, The Mammal Fauna of the Early Middle Pleistocene Cavern Infill Site of Westbury-sub-Mendip, Somerset. Special Papers in Palaeontology 28, 1982, 1-108.

Charles/Jacobi 1994: R. Charles / R. Jacobi, The Lateglacial Fauna from the Robin Hood Cave, Cresswell Crags: a Re-assessment. Oxford Journal of Archaeology 13/1, 1994, 1-32.

Clark 1925a: K. Clark, A Pioneer of Prehistory Part I. New Blackfriars 6/67, 1925, 603-613.

1925b: K. Clark, A Pioneer of Prehistory Part II. New Blackfriars 6/68, 1925, 640-648.

Dawkins 1874a: W. B. Dawkins, Cave Hunting. Researches on the Evidence of Caves Respecting the Early Inhabitants of Europe (London 1874).

1874b: W. B. Dawkins, The Probable Age of the *Machairodus* of Kent's Cavern. In: Dawkins 1874a, 330-335.

1877: W. B. Dawkins, On the Mammal-fauna of the Caves of Cresswell Crags. Quarterly Journal of the Geological Society 33, 1877, 589-612.

1903: W. B. Dawkins, On the Discovery of an Ossiferous Cavern of Pliocene Age at Doveholes, Buxton (Derbyshire). Quarterly Journal of the Geological Society 59, 1903, 105-132.

Dawkins/Sanford/Reynolds 1878: W. B. Dawkins / W. A. Sanford / S. H. Reynolds, A Monograph on the British Pleistocene Mammalia. 1: British Pleistocene Felidae. Palaeontographical Society 22 (London 1878).

Ewer 1954: R. F. Ewer, The Fossil Carnivores of the Transvaal Caves: Machairodontinae. Proceedings of the Zoological Society of London 125, 1954, 587-615.

Gervais 1848-1852: M. P. Gervais, Zoologie et paléontologie française, ed. A. Bertrand (Paris 1848-1852).

Jefferson/Tejada-Flores 1993: G. T. Jefferson / A. E. Tejada-Flores, The Late Pleistocene Record of *Homotherium* (Felidae: Machairodontinae) in the Southwestern United States. Palaeobios 15/3, 1993, 37-46.

Kennard 1945: A. S. Kennard, The Early Digs in Kent's Hole, Torquay, and Mrs. Cazalet. Proceedings of the Geologists' Association 1945, 156-192.

von Koenigswald 1934: G. H. R. von Koenigswald, Zur Stratigraphie des javanischen Pleistocän. De Ingenieur in Nederlandsch Indië 1934, 185-201.

Lankester 1869: E. R. Lankester, On the occurrence of *Machairodus* in the forest-bed of Norfolk. Geological Magazine 6, 1869, 440-441.

Lewis/Pacher/Turner 2010: M. E. Lewis / M. Pacher / A Turner, The larger Carnivora of the West Runton Freshwater Bed. Quaternary International 228, 2010, 116-135.

Lundberg/McFarlane 2007: J. Lundberg / D. A. McFarlane, Pleistocene Depositional History in a Periglacial Terrane: A 500 k.y. Record from Kents Cavern, Devon, United Kingdom. Geosphere 3/4, 2007, 199-219.

Lydekker 1885: R. Lydekker, Catalogue of the Fossil Mammalia in the British Museum (Natural History), Part 1 (London 1885).

Lyell 1873: C. Lyell, The Geological Evidences of the Antiquity of Man: with an Outline of Glacial and Post-tertiary Geology and Remarks on the Origin of Species, with Special Reference to Man's First Appearance on the Earth (London 1873).

MacEnery 1859: J. MacEnery, Cavern Researches, or, Discoveries of Organic Remains, and of British and Roman Reliques, in the Caves of Kents Hole, Anstis cove, Chudleigh, and Berry Head (London 1859).

McFarlane/Lundberg 2013: D. A. McFarlane / J. Lundberg, On the Occurrence of the Scimitar-toothed Cat, *Homotherium latidens* (Carnivora; Felidae), at Kents Cavern, England. Journal of Archaeological Science 40, 2013, 1629-1635.

Mello 1877: J. M. Mello, The Bone-caves of Creswell Crags. 3rd Paper. Quarterly Journal of the Geological Society 33, 1877, 579-588.

1879-1880: J. M. Mello, Notes on the More Recent Discoveries in the Cresswell Caves. Transactions of the Manchester Geological Society 15, 1879-1880, 290.

Merriam/Stock 1932: J. C. Merriam / C. Stock, The Felidae of Rancho La Brea. Carnegie Institute of Washington: Publications 422 (Washington 1932).

Mones/Rinderknecht 2004: A. Mones / A. Rinderknecht, The First South American Homotheriini (Mammalia: Carnivora: Felidae). Comunicaciones Paleontologicas II/35, 2004, 201-212.

Oakley 1980: K. P. Oakley, Relative Dating of the Fossil Hominids of Europe. Bulletin of the British Museum Natural History, Geology Series 34, 1980, 1-63.

Owen 1846: R. Owen, A history of British Fossil Mammals and Birds (London 1846).

Pengelly 1868: W. Pengelly, The Literature of Kent's Cavern, Torquay, prior to 1859. Reports and Transactions of the Devonshire Association for the Advancement of Science, Literature and Art 2, 2, 1868, 469-522.

1869: W. Pengelly, On the Alleged Occurrence of *Hippopotamus major* and *Machairodus latidens* in Kents Cavern, Torquay. Reports and Transactions of the Devonshire Association for the Advancement of Science, Literature and Art 3, 1869, 483-494.

1872: W. Pengelly, Notes on the *Machairodus latidens* Found by the Rev. J. MacEnery in Kent's Cavern. Reports and Transactions of the Devonshire Association for the Advancement of Science, Literature and Art 5, 1872, 165-179.

1873a: W. Pengelly, Notes on *Machairodus latidens* Found by the Rev. J. MacEnery in Kent's Cavern, Torquay. In: Report of the Forty-Second Meeting of the British Association for the Advancement of Science held at Brighton in August 1872 (London 1873) 119-120.

1873b: W. Pengelly, On the Kents Hole *Machairodus*. Quarterly Journal of Science III, 1873, 204-223.

1873c: W. Pengelly, Eighth Report of the Committee for Exploring Kent's Cavern, Devonshire. In: Report of the Forty-Second Meeting of the British Association for the Advancement of Science held at Brighton in August 1872 (London 1873) 28-47.

1875: W. Pengelly, Notes on a Tooth of *Machairodus latidens* in the Albert Memorial Museum. Reports and Transactions of the Devonshire Association for the Advancement of Science, Literature and Art 7, 1875, 247-260.

Rawn-Schatzinger 1992: V. Rawn-Schatzinger, The Scimitar Cat *Homotherium serum* Cope: Osteology, Functional Morphology, and Predatory Behaviour. Illinois State Museum Reports of Investigations 47, 1992, 1-80.

Reumer et al. 2003: J. W. F. Reumer / L. Rook / K. van der Borg / K. Post / D. Mol / J. de Vos, Late Pleistocene Survival of the Saber-toothed Cat *Homotherium* in Northwestern Europe. Journal of Vertebrate Paleontology 23/1, 2003, 260-262.

Rincón/Prevosti/Parra 2011: A. D. Rincón / F. J. Prevosti / G. E. Parra, New Saber-toothed Cat Records (Felidae: Machairodontinae) for the Pleistocene of Venezuela, and the Great American Biotic Interchange. Journal of Vertebrate Paleontology 31/2, 2011, 468-478.

Spencer/Melville 1974: H. E. P. Spencer / R. V. Melville, The Pleistocene Mammalian Fauna of Dove Holes, Derbyshire. Bulletin of the Geological Survey of Great Britain 48, 1974, 43-53.

Stuart 1974: A. J. Stuart, Pleistocene History of the British Vertebrate Fauna. Biological Reviews 49, 1974, 225-266.

Teilhard de Chardin 1938: P. Teilhard de Chardin, On Two Skulls of Machairodus from the Lower Pleistocene Beds of Choukoutien. Bulletin of the Geological Society of China 19/3, 1938, 235-256.

Turner 1999: A. Turner, Larger Carnivores (Mammalia, Carnivora) from Westbury Cave. In: P. Andrews / J. Cook / A. Currant / C. Stringer (eds), Westbury Cave: The Natural History Museum Excavations 1976-1984 (Bristol 1999) 175-193.

Turner/Antón 1997: A. Turner / M. Antón, The Big Cats and their Fossil Relatives (New York 1997).

SUMMARY / ZUSAMMENFASSUNG

A Very British Sabretooth: *Homotherium* in the UK

The sabretooth genus *Homotherium*, while wide-ranging in Eurasia, has not often been considered a critical component of the British Pleistocene fauna. The Early Pleistocene *Homotherium crenatidens* and the Middle to Late Pleistocene *Homotherium latidens* have, however, held a prominent position in 19th-century discussions of British palaeontology. As many of the UK sites containing *Homotherium* were first investigated over a century ago, the literature is confused on the question of what was found. An attempt has been made to track down all specimens of British *Homotherium* discussed in the literature and to give their current location. For the first time, the correct identity of all the *Homotherium* canines from Kent's Cavern in Devon is given and definitively associated with the famous Plate F of MacEnery. The history of the single *Homotherium* incisor found by MacEnery is deduced from study of his surviving writing and images.

Eine sehr britische Säbelzahnkatze: *Homotherium* im Vereinigten Königreich

Obwohl die Säbelzahngattung *Homotherium* in Eurasien weit verbreitet ist, wurde sie oft nicht als wichtiger Bestandteil der britischen pleistozänen Fauna angesehen. Das *Homotherium crenatidens* aus dem frühen Pleistozän und das *Homotherium latidens* aus dem mittleren bis späten Pleistozän nahmen jedoch in den Diskussionen über die britische Paläontologie des 19. Jahrhunderts eine herausragende Stellung ein. Da viele der britischen Fundorte von *Homotherium* vor mehr als einem Jahrhundert erstmals untersucht wurden, ist die Literatur in der Frage, was gefunden wurde, unübersichtlich. Es wurde versucht, alle in der Literatur besprochenen Exemplare des britischen *Homotherium* aufzuspüren und ihren aktuellen Fundort anzugeben. Zum ersten Mal wird die korrekte Identität aller *Homotherium*-Eckzähne aus der Kent's Cavern in Devon angegeben und definitiv mit der berühmten Platte F von MacEnery in Verbindung gebracht. Die Geschichte des einzelnen *Homotherium*-Schneidezahns, der von MacEnery gefunden wurde, wird aus dem Studium seiner überlieferten Schriften und Bilder abgeleitet.

no.	element	level	museum	accession	figured	notes	references
1	left I/1 or I/2	Westbury Unit 18 (Bed 3)	British Museum Natural History (NHM)	M47488			Turner 1999
2	left M/1	Westbury Unit 19/8	British Museum Natural History (NHM)	ME-50 (2-3 ME50)	Turner 1999, fig. 8.1B		Turner 1999
3	left canine	Westbury Unit 19/14	British Museum Natural History (NHM)	1982-43			Turner 1999
4	right upper canine	Westbury Bed 4a	British Museum Natural History (NHM)	M33983	Turner 1999, fig. 8.1A Bishop 1982, pl. 4 fig. 9		Turner 1999 Bishop 1982
5	lower canine	Westbury unstratified		M33984		same as 6?	Bishop 1982
6	canine	Westbury unstratified	British Museum Natural History (NHM)	F48		same as 5?	Turner 1999
7	M/1	Westbury unstratified	British Museum Natural History (NHM)	M33703	Turner 1999, fig. 8.1C Bishop 1982, pl. 4 fig. 8		Turner 1999 Bishop 1982
8	upper canine fragment	Westbury unstratified	British Museum Natural History (NHM)	M33700a			Bishop 1982
9	upper canine fragment	Westbury unstratified	British Museum Natural History (NHM)	M33700b			Bishop 1982
10	upper incisor	Westbury unstratified	British Museum Natural History (NHM)	M33705			Bishop 1982
11	upper P4 fragment	Westbury unstratified	British Museum Natural History (NHM)	M33702			Bishop 1982
12	lower canine	Westbury unstratified	British Museum Natural History (NHM)	M33701a			Bishop 1982
13	lower canine	Westbury unstratified	British Museum Natural History (NHM)	M33701b			Bishop 1982

Tab. 4 The Westbury-sub-Mendip material.

no.	element	site	museum	accession	figured	notes	references
1	right mandible	Kessingland/ Pakefield	Natural History Division, National Museum of Ireland, Dublin	NMING : F15001 (NMINH 1907.326)	Backhouse 1886, pl. 10	1907 Backhouse Sale to Dublin Museum (Special Correspondence 1907)	Backhouse 1886 Kennard 1945 Stuart 1982
2	?	Kessingland/ Pakefield	British Museum of Natural History (NHM)	?		Mutch, Stewart and Durbridge Private Collection 1994-1998	Stuart 2001
3	?	Kessingland/ Pakefield	British Museum of Natural History (NHM)	?		Mutch, Stewart and Durbridge Private Collection 1994-1998	Stuart 2001
4	right upper canine	Cromer	?	?	Lankester 1869, pl. 16 Newton 1882, pl. 4 fig. 5. 5a	Mr. Jarvis of Cromer → A.Savin possible same as 8	Lankester 1869 Newton 1882 Backhouse 1886
5	incisor	Cromer	?	?		in possession of E. T. Newton	Backhouse 1886
6	?	Covehithe	?	?			Stuart 1974
7	left upper canine	Bacton	British Museum of Natural History (NHM)	M17926			Stuart 1974 Turner 2009 Turner 1999
8	fragmental right upper canine crown	Sidestrand	British Museum Natural History (NHM)	M6084		possibly same as 4	Stuart 1974 Turner 2009
9	right third metatarsal	West Runton, Freshwater Bed	British Museum Natural History (NHM)	M17896			Turner 2009 Turner 1999 Lewis/Pacher/Turner 2010
10	left calcaneum	West Runton Freshwater Bed	British Museum Natural History (NHM)	M17903	Lewis/Pacher/Turner 2010, fig. 8		Lewis/Pacher/Turner 2010
11	carnassial M1	Thorpe St. Andrews	Norwich Castle Museum	NWHCM: 1894.76.2006:G	Newton 1891, pl. 1 fig. 2a-b	fitch donated to Norwich Castle Museum 1894	Newton 1891 Cleevely 1983
12	left humerus	Cromer Forest Bed, Norfolk	Norwich Castle Museum	NWHCM: 1898.51.FC373:G		Colman Collection – probably lion (*P. fossilis*/*P. spelaea*) (A. Turner pers. comm.)	NMAS Online 2010

Tab. 5 The East Anglian material.

EXTINCTION

ULRICH JOGER

ICE AGE FAUNA IN THE BRAUNSCHWEIG REGION, WITH A FOCUS ON THE CARNIVORES

Braunschweig used to be an independent duchy for more than 1000 years. It included territories at the River Weser in the West as well as territories which are now part of Saxony-Anhalt in the East. Parts of the geologically important Harz Mountains used to be territory of Braunschweig, too. However, the territory as a whole was not contiguous but consisted of many isolated spots which were separated by Hanoverian or Prussian territory in between (**fig. 1**). Therefore, it appears logical to define the »Braunschweig region« as the area which encompasses all these separate Brunsvigian parts, as well as the area in between. Nowadays, the central and eastern parts of the area form part of Germany's largest National Geopark »Harz – Braunschweiger Land – Ostfalen« which is also recognized as a UNESCO Geopark.

Geologically the area is rich in Mesozoic sediments (Triassic, Jurassic, and Cretaceous), and important fossil sites such as Cremlingen (Lkr. Wolfenbüttel/D) and Langenberg (Lkr. Gütersloh/D) have yielded vertebrate fossils such as ichthyosaurs, pterosaurs and dinosaurs (Fischer et al. 2012; Hauff et al. 2014; Sander et al. 2006).

As far as Pleistocene localities are concerned, the majority of them are Late Pleistocene, but three important Middle Pleistocene localities are also known (**tab. 1**).

Rübeland Caves – 10 000-45 000 BP	
Salzgitter-Lebenstedt – 55 000-50 000 BP	
Salzgitter-Thiede (Gypsum)	Upper Pleistocene
Peine-Förste (Gypsum)	Upper Pleistocene
Westeregeln	Upper Pleistocene
Walkenried (Gypsum)	Upper Pleistocene
Osterode-Düna	Upper Pleistocene
Unicorn Cave Scharzfeld – 35 000-170 000 BP	
Lehringen near Verden (out of area) – 124 000 BP	
Gröbern (out of area) – 120 000-125 000 BP	
Neumark-Nord (out of area) – *c.* 200 000 BP	
Schöningen – *c.* 300 000 BP	
Bilzingsleben (out of area) – 350 000-420 000 BP	

Tab. 1 Principle Pleistocene localities in the Braunschweig region (dating of Middle Pleistocene localities under debate; see text).

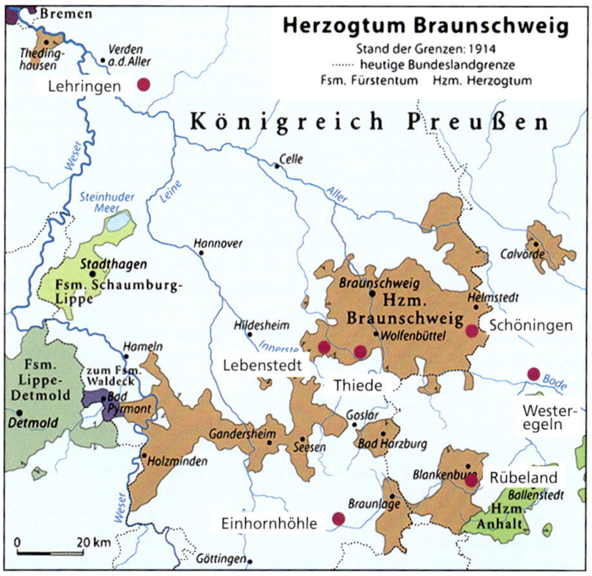

Fig. 1 Historical borders of the Duchy of Braunschweig and localities mentioned. – (SNHM Archives, modified).

In: Nicholas J. Conard · Henning Hassmann · Kurt Felix Hillgruber · Jordi Serangeli · Thomas Terberger (eds), The *Homotherium* Finds from Schöningen 13II-4. Man and Big Cats of the Ice Age. Forschungen zur Urgeschichte aus dem Tagebau von Schöningen 4 (Mainz 2022). DOI: https://doi.org/10.11588/propylaeum.1006.c13522

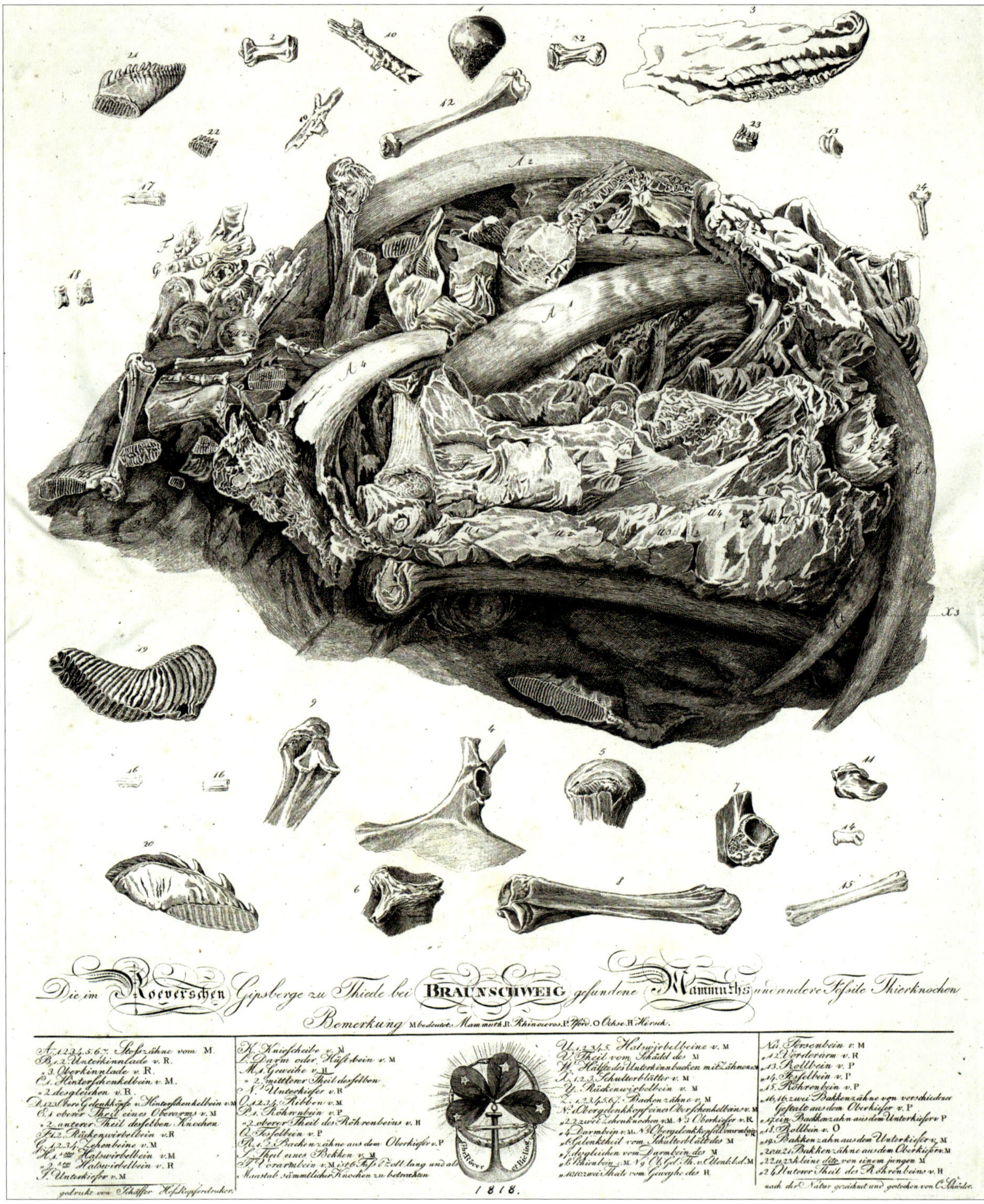

Fig. 2 Historical sketch of a fossil assemblage from a gypsum layer at Salzgitter-Thiede (1818). – (Courtesy University Library Göttingen).

LATE PLEISTOCENE LOCALITIES

Fossil bones from the Late Pleistocene are numerous in the Braunschweig region. Historically, four localities stand out, as they have been known and documented for centuries: Bone materials from the first two localities are stored in the State Natural History Museum (SNHM) in Braunschweig.

Fig. 3 Entrance of Baumann's Cave. – (After C. Bruno in: Merian 1654; Collection Stephan Kempe).

Salzgitter-Thiede

In his posthumously published book »Protogaea« (1749), the famous G. W. Leibniz (1646-1716) depicted a molar of a mammoth from »Tida« (now Salzgitter-Thiede) which he interpreted as a tooth of a marine animal. Historical mammoth molars from that locality (a gypsum quarry) have been deposited in the Geoscience Museum of the University of Göttingen and in the State Natural History Museum in Braunschweig. The latter were recovered in the years 1810-1818 by Carl Bieling and J. G. J. Ballenstedt (**fig. 2**). Ballenstedt was among the first German palaeontologists. In 1817 he published a book called »Beweis von dem Daseyn und Untergange von mehr als einer Vorwelt« (Joger 2005).

The fossils from Thiede have not yet been radiocarbon dated, but they are probably Late Pleistocene in age. Apart from woolly mammoth (*Mammuthus primigenius*) and diverse small mammals (Nehring 1878), they comprise of woolly rhinoceros (*Coelodonta antiquitatis*), horse (*Equus* sp.), reindeer (*Rangifer tarandus*), deer (*Cervus* sp.), steppe bison (*Bison priscus*), badger (*Meles meles*), polecat (*Mustela putorius*), polar fox (*Alopex lagopus*), wolf (*Canis lupus*), cave lion (*Panthera leo spelaea*), and hyena (*Crocuta crocuta spelaea*), respectively their bite marks on other bones. Diedrich (2007) classifies the bone assemblages at Thiede, Osterode (Lkr. Göttingen/D), and Westeregeln (Salzlandkreis/D) near Magdeburg as hyena dens, and he also reports that hyenas were frequently scavenging on lion bones (Diedrich 2011).

The Rübeland Caves

Baumann's and Hermann's Caves, also known as the Rübeland Caves, are situated in the Bode Valley near Wernigerode (Lkr. Harz/D) in Saxony-Anhalt. Historically, they were part of the Blankenburg exclave of the Duchy of Braunschweig. The caves are situated in Middle Devonian limestone and the formation of the caves is thought to have started in the Neogene. In connection with the Neogene-Quaternary development of the Bode River Valley, many vertically connected horizontal levels emerged. The latest and deepest is a phreatic zone in today's karst water level.

Fig. 4 Wilhelm Blasius (1845-1912), director of the State National History Museum in Brunswick. – (SNHM Archives).

Fig. 5 Fossil-rich fluviatile sediment (unstratified) in Hermann's Cave. – (SNHM Archives).

First mentioning of speleothem formations and bone finds in Baumann's Cave dates back to the 16th century (Kempe et al. 2004). Due to the richness in bones, then interpreted as remains of the unicorn *Unicornu fossile*, the cave gained the attention of numerous historical authors, for example G. Agricola (1546), C. Gessner (1565), M. Merian (1654; **fig. 3**), G. W. Leibniz (1749), and C. Linnaeus (1779). Already in 1665 the first map of Baumann's Cave was published. The stalagmites, but also bones and teeth of »enormous and terrible animals« or »giants« found in that cave were harvested and sold as remedies. To regulate that trade, Duke Rudolf August of Braunschweig and Lüneburg issued the world's first cave protection order for this cave in 1668, and employed the first official cave warden. In the 18th century, bones from the cave were still sold as remnants of the mythical unicorn. J. F. Zückert (1763) interpreted them as relicts of the great flood of the bible. In 1789 the first correct interpretation of the fossil bones found as remains of bears was published (Lasius 1789). This was a few years before the formal description of the cave bear as *Ursus spelaeus.* Yet even William Buckland (1824) still used Baumann's Cave as an example for the great flood.

Survey works in 1888 discovered important and bone-rich new parts of the cave. Thereafter, the State National History Museum in Brunswick conducted palaeontological excavations from 1889 until 1902.

Hermann's Cave was found during road construction works in 1866. First palaeontological excavations took place in 1875 (Grotrian 1878) and continued until 1897 by the Technical University of Braunschweig and the State National History Museum in Brunswick under J. H. Kloos and W. Blasius (**fig. 4**). Kloos and Müller (1889) published a description of the caves with a set of high quality photographs of the caves' interior and their sediments (**fig. 5**). Further excavations by the State Museum of Prehistory in Halle took place in 1954 and 1962, as well as by the Karstmuseum in Uftrungen (Völker 1998).

The caves' fauna is highly diverse. Bird and mammal bones dominate, the majority of the latter belong to cave bears (*Ursus spelaeus*). The presence of cave lion (*Panthera leo spelaea*), leopard (*Panthera pardus*; **fig. 6**), cave hyena (*Crocuta crocuta spelaea*), wolf (*Canis lupus*), red fox (*Vulpes vulpes*), arctic fox (*Alopex lagopus*), marten (*Martes* sp.), stoat (*Mustela erminea*), wolverine (*Gulo gulo*), and wildcat (*Felis silvestris*) is also documented; presumed skull material of dhole (*Cuon alpinus*) is under study.

Fig. 6 A well-preserved leopard humerus from Baumann's Cave at Rübeland. – (Courtesy W. Rosendahl, Mannheim).

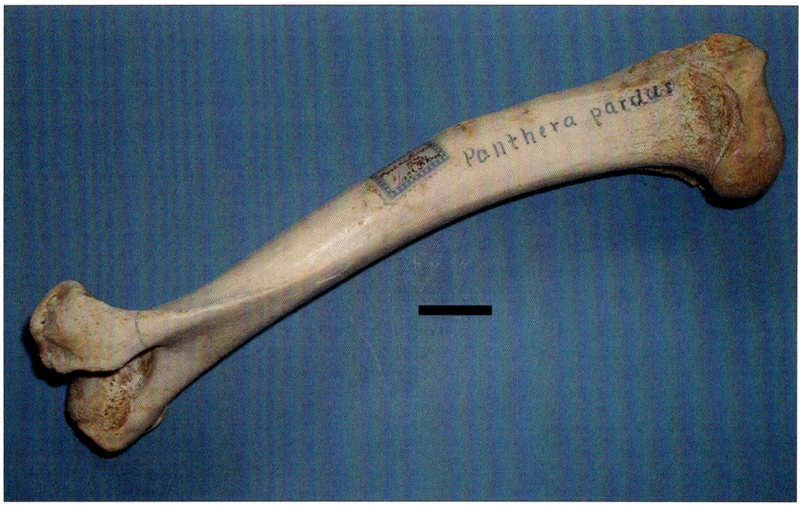

The oldest radiocarbon datings (44-46 ka uncalibrated) were obtained in *Ursus spelaeus* and *Rangifer tarandus*, whereas bones of *Canis lupus*, *Panthera pardus* and *P. leo spelaea* were dated between 31 and 37 ka (Joger/Rosendahl 2012). This period (OIS 3) is characterized by frequent oscillations of mean temperature, and even interstadial-like »warm« phases (Dansgaard et al. 1993). Only a chamois (*Rupicapra rupicapra*) was estimated uppermost Pleistocene (10218 ± 71 years BP) (Joger/Rosendahl 2012).

Ecologically, the fauna includes five different elements: arctic fauna (e. g. *Alopex lagopus*, *Rangifer tarandus*, *Ovibos moschatus*, *Gulo gulo*); steppe fauna (e. g. *Allactaga major*, *Bison priscus*); alpine fauna (e. g. *Capra ibex*, *Rupicapra rupicapra*); temperate fauna (e. g. *Capreolus capreolus*, *Cervus elaphus*, *Vulpes vulpes*) and widespread carnivores (e. g. *Canis lupus*, *Panthera leo spelaea*, *P. pardus*, *Crocuta crocuta spelaea*). It is uncertain whether all of these species existed contemporaneously or in subsequent periods.

Blasius (1898) reported Neanderthal-type artefacts from Baumann's Cave. Among them were stone tools, but also a number of broken and modified bones. In his view, they had been handled by humans. Yet some of these bones could have been chewed by hyenas.

The Unicorn Cave (**Einhornhöhle**) near Scharzfeld/Harz (Lkr. Göttingen/D) is peculiar in preserving sediments representing an enormous time span, ranging from 10000 to 200000 years of age, hence incorporating both Late and Middle Pleistocene (Saale complex, Eem). However, although the cave has been known for a long time (Agricola 1546; Buckland 1824), there have only been small-scaled excavations (profiles of up to 14 m in depth). Carnivore bones found represent *Ursus spelaeus*, *Panthera leo spelaea*, *Felis sylvestris, Canis lupus, Crocuta crocuta spelaea, Meles meles,* and *Gulo gulo* (Nielbock 2010). The Middle Pleistocene (lower) layers in the cave are yet to be uncovered.

Another Late Pleistocene locality, **Salzgitter-Lebenstedt**, has become famous because of the Neanderthal skull elements found there in 1952 and in 1976/1977, together with stone tools, worked bones and antlers, and remains of the following megafauna: *Panthera leo spelaea*, *P. pardus* (not confirmed), *Canis lupus, Mammuthus primigenius, Coelodonta antiquitatis, Bison priscus, Rangifer tarandus, Equus* sp., *Megaloceros giganteus* (Grote/Thieme 1985). The assemblage is dated to the interstadial of Oerel (59000-55000 years BP); it is interpreted as a Neanderthal campsite (Steinmetz 2005a). A botanical analysis revealed a cool-temperate herbal steppe with dwarfed *Betulus*, *Juniperus* and *Salix*, in a river valley with some semi-aquatic vegetation. Finds of a Russian Jerboa (*Allactaga major*) indicate a steppic vegetation. Most finds from Salzgitter-Lebenstedt are stored in the Braunschweigisches Landesmuseum (see Ludovici/Pöppelmann 2017).

MIDDLE PLEISTOCENE LOCALITIES

Lehringen (between Verden and Bremen [Lkr. Verden/D]) is far to the West of our region, but this interglacial locality (Eem, pollen zone IIIc, dated 124 000 years BP) represents another Middle Pleistocene hunting site in Lower Saxony with well-preserved wooden tools. Bones of a forest elephant (*Palaeoloxodon antiquus*) with a wooden lancet among them were found at Lehringen in 1948 (Adam 1951). With it, bones of *Ursus* cf. *arctos*, *Canis lupus*, *Lutra lutra*, *Felis sylvestris*, *Bos primigenius*, *Capreolus capreolus*, *Megaloceros giganteus*, *Dama dama*, *Cervus elaphus*, *Equus* sp., and *Stephanorhinus hemitoechus* were recovered (Houben 2003). This is a typical interglacial fauna. The warm temperate climate is also evidenced by a complete shell of a pond turtle, *Emys orbicularis*. The locality was probably a lakeshore, comparable with the older Schöningen site or with Neumark-Nord, which may be roughly contemporaneous (see below).

Gröbern (near Bitterfeld in Saxony-Anhalt [Lkr. Anhalt-Bitterfeld/D]) to the Southeast of our region (pollen zone IIIb) was nearly contemporaneous with Lehringen. An adult *Palaeoloxodon antiquus* had been butchered there by early Neanderthals (Steinmetz 2005b). Bones of *Dama dama* and *Capreolus capreolus* were also found at Gröbern, but no carnivore remains.

The nearby site **Neumark-Nord** (Saalekreis/D) revealed, according to Mania (2010), an interglacial fauna within the Saale complex (roughly 200 000 years BP), whereas tooth enamel ESR dating gave a younger, Eemian age (Schüler 2010). *Palaeoloxodon antiquus* was abundant at Neumark-Nord, but a nearly complete skeleton of a lioness (*Panthera leo spelaea*) was also found (Diedrich 2010; Fischer 2010). Good overviews and detailed descriptions of the locality and its fauna are given in the recent publications edited by Meller (2010a; 2010b).

Like Gröbern and Neumark-Nord, the famous hominid locality **Bilzingsleben** (Lkr. Sömmerda/D) lies outside the Braunschweig area which is the scope of this article, but it is of similar age or even slightly older than Schöningen. The absolute age is estimated between 420 000-350 000 years BP (Mania/Mai 2001), but an alternative dating ranges between only 250 000-200 000 years BP (Eissmann/Litt 1994). Among the human remains belonging to several individuals, bones of *Palaeoloxodon antiquus*, *Stephanorhinus kirchbergensis/hemitoechus*, *Bos primigenius*, *Bison* sp., *Equus mosbachensis*, *Capreolus suessenbornensis*, *Megaloceros* sp., *Dama dama*, *Cervus elaphus*, *Sus scrofa*, *Ursus* cf. *deningeri*, *Panthera leo spelaea*, *Felis silvestris*, *Vulpes vulpes*, *Meles meles*, *Martes* sp., *Lutra lutra*, *Crocuta crocuta spelaea*, *Canis lupus*, the primate *Macaca floreana*, and the beaver *Trogontherium cuvieri* were found (Müller/Pasda 2011). Fauna and flora are evidence for a comparatively warm climate. Oak forests were dominant.

Schöningen (Lkr. Helmstedt/D) – like Neumark-Nord an Eocene open-air lignite mine with a series of Pleistocene layers filling a depression on top – is the focus of this volume and a comprehensive list of the large mammalian fauna secured there so far shall therefore be enough for this overview. The exact age and stratigraphy have been under debate until recently. Current thermoluminescence dating has reduced the estimated age of the Middle Pleistocene »spear horizon« from 400 000 to approximately 300 000 years BP (MIS 9; Richter/Krbetschek 2015).

For the mammal faunal assemblages, I am referring here to van Kolfschoten et al. (2007), as well as van Kolfschoten (2014), who differentiate three main strata:

	early Middle Pleistocene 780 000-300 000 years BP	late Middle Pleistocene 300 000-126 000 years BP	Late Pleistocene 126 000-11 700 years BP
lion	*Panthera leo fossilis* (= *P. mosbachensis*)	*Panthera leo spelaea* (since 370 000 BP)	*Panthera leo spelaea*
leopard	*Panthera pardus sickenbergi*	*Panthera pardus antiqua*	*Panthera pardus spelaea*
jaguar	*Panthera onca gombaszoe-gensis*		
sabre-toothed cat	*Homotherium crenatidens*	*Homotherium latidens*	*Homotherium latidens* (?) – North Sea specimen
lynx	*Lynx issiodorensis*	*Lynx issiodorensis*	*Lynx lynx*
cheetah	*Acinonyx pardinensis*		
cougar	*Puma pardoides* (= *Panthera schaubi*)		
hyena	*Crocuta crocuta* ssp.	*Crocuta crocuta spelaea*	*Crocuta crocuta spelaea*

Tab. 2 Fossil Felidae in Central Europe (including Hyaenidae).

- Schöningen 12II (»Reinsdorf-Interglacial«, MIS 9): *Palaeoloxodon antiquus, Dicerorhinus (Stephanorhinus) kirchbergensis, Bos primigenius, Bison* sp., *Equus mosbachensis, Capreolus capreolus, Cervus elaphus, Sus scrofa, Ursus spelaeus, U. thibetanus, Martes* sp., *Crocuta crocuta spelaea, Canis lupus.*
- Schöningen 13I (probably the oldest stratum; may be »Holstein-Interglacial«, MIS 11: *Mammuthus trogontherii* (probably misidentified *Palaeoloxodon antiquus*), *Equus mosbachensis, Cervus elaphus, Bison* sp.
- Schöningen 13II (»Reinsdorf-Interglacial«, MIS 9): *Palaeoloxodon antiquus, Bos primigenius, Bison priscus, Equus mosbachensis, Cervus elaphus, Megaloceros giganteus, Canis lupus, Vulpes vulpes, Mustela erminea, M. nivalis.* Among the smaller mammals, bones of a Desman (*Desmana* sp.) are a very remarkable find (van Kolfschoten/van Asperen/Voormolen 2007). The list was recently complimented by the rhinoceros species *Stephanorhinus kirchbergensis* and *S. hemitoechus*, by *Equus hydruntinus* and by the saber-toothed cat *Homotherium latidens* (van Kolfschoten/Buhrs/Verheijen 2015).

The Holsteinian fauna of Schöningen 13I is considered to be older (MIS 11 according to van Kolfschoten 2014) than the »spear horizon« (MIS 9), but thermoliminescence data (Richter/Krbetschek 2015) resulted in just a slightly older age for Schöningen 13I.

The faunal difference between Schöningen 12II (with forest species like *Capreolus capreolus* and *Sus scrofa*) as well as the thermophilic *Bubalus murrensis* (Serangeli et al. 2015a) on one side, and Schöningen 13II (with a dominance of steppe species) on the other side, is interpreted as an effect of climatic cooling.

However, Heinrich/van Kolfschoten (2007) also report the tertiary relict species *Trogontherium cuvieri* from Schöningen 13II, pointing to a temperate climate with trees. Another, much more spectacular tertiary relict found was the saber-toothed cat *Homotherium latidens* (Serangeli et al. 2015b).

DISTRIBUTION AND ABUNDANCE OF CARNIVORES

While the smaller carnivores (Mustelidae, foxes, lynx and wildcat) are represented in Pleistocene localities of the region with the same species as today, there is a marked faunal change within the large mammals. For example, the cave bear was present in the Harz Mountains in great numbers until about 30 000 years BP, but died out soon afterwards. As the Middle Pleistocene localities reported here are not caves (with exception of deep layers of the Einhornhöhle), it is not surprising that normally no cave bears were found there. Yet among the bones from Schöningen 12II, with a more forested environment than Schöningen 13II, even two bear species (*Ursus spelaeus* und *U. thibetanus*) could be distinguished (van Kolfschoten 2014).

Three large carnivores are abundant in most Pleistocene localities: Lion (*Panthera leo*), wolf (*Canis lupus*) and hyena (*Crocuta crocuta*). The latter is often represented in the fossil record – not necessarily by its own bones, but rather by its characteristic bitemarks, engraved in other species' bones, or by traces left by its habit of chewing at the end of those bones. Some Late Pleistocene localities are even interpreted as hyena dens – the bones found there were accumulated by foraging hyenas (Diedrich 2007). Much less abundant is the leopard (*Panthera pardus*), and the leopard bones from Rübeland and Salzgitter-Lebenstedt are amongst the northernmost records of this species in the world (Probst 2011; Paijmans et al. 2018). This fact probably indicates that leopards were absent from our region during cold periods of the Pleistocene.

Cheetah (*Acinonyx* sp.), cougar (*Puma pardoides*) and jaguar (*Panthera onca gombaszoegensis*) were present in the early Middle Pleistocene Mosbach sands near Wiesbaden as well as at Untermaßfeld (Lkr. Schmalkalden-Meiningen/D) in Thuringia (Hemmer 1971; Hemmer/Schütt 1970; Hemmer et al. 2003). Up to now, they have not been recorded from sites younger than the early Middle Pleistocene (**tab. 2**). However, as witnessed by the discovery of saber-toothed cat remains (*Homotherium* was thought to have become extinct around the border between early and late Middle Pleistocene), an extension of a stratigraphic range of a species is always possible. Only the isolated North Sea specimen of *Homotherium latidens* (see contribution Mol and Langeveld in this volume) was dated younger than the one from Schöningen (even Late Pleistocene, but lacking any stratigraphic context).

REFERENCES

Adam 1951: L. D. Adam, Der Waldelefant von Lehringen – eine Jagdbeute des diluvialen Menschen. Quartär 5, 1951, 79-92.

Agricola 1546: G. Agricola, De natura fossilium. De ortu et causis subterraneorum (Basel 1546).

Blasius 1898: W. Blasius, Die anthropologisch wichtigen Funde in den Höhlen bei Rübeland a/H. Correspondenz-Blatt der Deutschen anthropologischen Gesellschaft 10, 1898, 109-113.

Buckland 1824: W. Buckland, Reliquiae Diluvianae; or, Observations on the organic remains contained in caves, fissures, and diluvial gravel, and on other geological phenomena, attesting the action of an universal deluge (London [2]1824).

Dansgaard et al. 1993: W. Dansgaard / S. J. Johnsen / H. B. Clausen / D. Dahl-Jensen / N. S. Gundestrup / C. U. Hammer / C. S. Hvidberg / J. P. Steffensen / A. E. Sveinbjörnsdottir / J. Jouzel / G. Bond, Evidence for general instability of past climate from a 250-kyr ice-core record. Nature 364, 1993, 218-220.

Diedrich 2007: C. G. Diedrich, The Upper Pleistocene *Crocuta crocuta spelaea* (Goldfuss 1823) population and its prey from the gypsum karst den site Westeregeln near Madgeburg (Middle Germany). Abhandlungen und Berichte für Natukunde 30, 2007, 57-83.

2010: C. G. Diedrich, A diseased *Panthera leo spelaea* (Goldfuss 1810) lioness from a forest elephant graveyard in the Late Pleistocene (Eemian) interglacial lake at Neumark-Nord, central Germany. Historical Biology Online 2010, 1-23.

2011: C. G. Diedrich, Late Pleistocene steppe lion *Panthera leo spelaea* (Goldfuss, 1810) footprints and bone records from open air sites in northern Germany – Evidence of hyena-lion antagonism and scavenging in Europe. Quaternary Science Reviews 30, 2011, 1883-1906.

Eissmann/Litt 1994: L. Eissmann / T. Litt (eds), Das Quartär Mitteldeutschlands. Ein Leitfaden und Exkursionsführer; mit einer Übersicht über das Präquartär des Saale-Elbe-Gebietes. Altenburger Naturwissenschaftliche Forschungen 7 (Altenburg 1994).

Fischer 2010: K. Fischer, Ein Löwenskelett (*Panthera spelaea*, GOLDFUSS, 1810) aus interglazialen Seesedimenten der Saalezeit von Neumark-Nord bei Merseburg. In: Meller 2010b, 339-360.

Fischer et al. 2012: V. Fischer / M. W. Maisch / D. Naish / R. Kosma / J. Liston / U. Joger / F. J. Krüger / J. Pardo Pérez / J. Tainsh / R. M. Appleby, New ophthalmosaurid ichthyosaurs from the European Lower Cretaceous demonstrate extensive ichthyosaur survival across the Jurassic-Cretaceous boundary. PLoS ONE 7/1, 2012, e29234. DOI: 10.1371/journal.pone.0029234.

Gessner 1565: C. Gessner, De Omni Rervm Fossilivm Genere, Gemmis, Lapidibvs, Metallis, Et Hvivsmodi (Zürich 1565).

Grote/Thieme 1985: K. Grote / H. Thieme, Eiszeitliche Jagdtiere und Jäger der mittleren Altsteinzeit am Beispiel der Freilandstation Salzgitter-Lebenstedt. In: K. Wilhelmi (ed.), Ausgrabungen in Niedersachsen. Archäologische Denkmalpflege 1979-1984. Berichte zur Denkmalpflege in Niedersachsen, Beiheft 1 (Stuttgart 1985) 51-57.

Hauff et al. 2014: R. B. Hauff / C. Heunisch / U. Hochsprung / J.-M. Ilger / U. Joger / M. Klopchar / R. Kosma / F.-J. Krüger / D. Thies / H. Zellmer, Jurameer. Niedersachsens versunkene Umwelt (München 2014).

Heinrich/van Kolfschoten 2007: W.-D. Heinrich / T. van Kolfschoten, Erster Skelettfund von *Trogontherium cuvieri* (Altbiber). In: H. Thieme (ed.), Die Schöninger Speere – Mensch und Jagd vor 400.000 Jahren (Stuttgart 2007) 116-123.

Hemmer 1971: H. Hemmer, Zur Kenntnis pleistozäner mitteleuropäischer Pantherkatzen (Pantherinae). Veröffentlichungen der Zoologischen Staatssammlung München 1, 1971, 15-36.

Hemmer/Schütt 1970: H. Hemmer / G. Schütt, Ein Gepardenfund aus den Mosbacher Sanden (Altpleistozän, Wiesbaden). Mainzer Naturwissenschaftliches Archiv 9, 1970, 118-131.

Hemmer et al. 2003: H. Hemmer / R. D. Kahlke / T. Keller, *Panthera onca gombaszoegensis* (KRETZOI, 1938) aus den frühmittelpleistozänen Mosbach-Sanden (Wiesbaden, Hessen, Deutschland). Ein Beitrag zur Kenntnis der Variabilität und Verbreitungsgeschichte des Jaguars. Neues Jahrbuch für Geologie und Paläontologie, Abhandlungen 229, 2003, 31-60.

Houben 2003: C. Houben, Die Wirbeltierfauna aus dem letzten Interglazial von Lehringen (Niedersachsen, Deutschland). Eiszeitalter und Gegenwart 52, 2003, 25-39.

Joger 2005: U. Joger, Johann Justus Ballenstedt und die Anfänge der Paläontologie. Braunschweigischer Kalender 2005, 98-99.

Joger/Rosendahl 2012: U. Joger / W. Rosendahl, The Rübeland Caves (Harz Mts.) – Historical excavations and modern analyses. Braunschweiger Naturkundliche Schriften 11, 2012, 55-68.

Kempe et al. 2004: S. Kempe / B. Dunsch / K. Fetkenheuer / G. Naumann / F. Reinboth, Die Baumannshöhle bei Rübeland/Harz im Spiegel der wissenschaftlichen Literatur vom 16. bis zum 18. Jahrhundert: Lateinische Quellentexte. Braunschweiger Naturkundliche Schriften 1/7, 2004, 171-215.

Kloos/Müller 1889: J. H. Kloos / M. Müller, Die Hermannshöhle bei Rübeland (Weimar 1889).

van Kolfschoten 2007: T. van Kolfschoten, Die Kleinsäugerreste aus dem Reinsdorf-Interglazial von Schöningen. In: H. Thieme (ed.), Die Schöninger Speere – Mensch und Jagd vor 400.000 Jahren (Stuttgart 2007) 112-115.

2014: T. van Kolfschoten, The Palaeolithic locality Schöningen (Germany): A review of the mammalian record. Quaternary International 326-327, 2014, 469-480.

van Kolfschoten/van Asperen/Voormolen 2007: T. van Kolfschoten / E. van Asperen / B. Voormolen, Die Großsäugerfauna von Schöningen. In: H. Thieme (ed.), Die Schöninger Speere – Mensch und Jagd vor 400.000 Jahren (Stuttgart 2007) 76-86.

van Kolfschoten/Buhrs/Verheijen 2015: T. van Kolfschoten / E. Buhrs / I. Verheijen, The larger mammal fauna from the Lower Paleolithic Schöningen spear site and its contribution to hominin subsistence. Journal of Human Evolution 89, 2015, 138-153.

Leibniz 1749: G. W. Leibniz, Protogaea (Göttingen 1749).

Linnaeus 1779: C. von Linnaeus/Linné, Vollständiges Natursystem des Mineralreichs; nach der 12. lat. Ausgabe in einer freyen und vermehrten Uebersetzung von J. F. Gmelin (Nürnberg 1779).

Lasius 1789: G. S. O. Lasius, Beobachtungen über die Harzgebirge, nebst einem Profilrisse, als ein Beytrag zur mineralogischen Naturkunde (Hannover 1789).

Ludovici/Pöppelmann 2017: B. Ludovici / H. Pöppelmann (eds), Die Tierknochenfunde der mittelpaläolithischen Jägerstation von Salzgitter-Lebenstedt. Forschungen und Berichte des Braunschweigischen Landesmuseums N. F. 1 (Braunschweig 2017).

Mania 2010: D. Mania, Zur Einordnung der Warmzeit von Neumark-Nord und ihrer Elefanten-Fauna in den Ablauf der Erdgeschichte. In: Meller 2010a, 64-70.

Mania/Mai 2001: D. Mania / H.-D. Mai, Molluskenfaunen und Floren im Elbe-Saalegebiet während des mittleren Eiszeitalters. Praehistoria Thuringica 6/7, 2001, 46-91.

Meller 2010a: H. Meller (ed.), Elefantenreich – eine Fossilwelt in Europa [exhibition catalogue] (Halle [Saale] 2010).

2010b: H. Meller (ed.), Neumark-Nord – Ein interglaziales Ökosystem des mittelpaläolithischen Menschen. Veröffentlichungen des Landesamtes für Denkmalpflege und Archäologie Sachsen-Anhalt. Landesmuseum für Vorgeschichte 62 (Halle [Saale] 2010).

Merian 1654: M. Merian, Topographia und eigentliche Beschreibung der vornembsten Stäte, Schlösser auch andere Plätze und Örter in den Herzogthümern Braunschweig und Lüneburg, und denen dazu gehörenden Grafschafften und Landen (Frankfurt 1654).

Müller/Pasda 2011: W. Müller / C. Pasda, Site formation and faunal remains of the Middle Pleistocene site Bilzingsleben. Quartär 58, 2011, 25-49.

Nehring 1878: A. Nehring, Die quaternen Faunen von Thiede und Westeregeln nebst Spuren des vorgeschichtlichen Menschen II. Archiv des Anthropologischen Organs der Deutschen Geselschaft für Anthropologie, Ethnologie und Urgeschichte 11, 1878, 1-64.

Nielbock 2010: R. Nielbock, Die Einhornhöhle. Die Welt der Einhörner, Höhlenbären und Neandertaler (München 2010).

Paijmans et al. 2018: J. L. A. Paymans / A. Barlow / D. W. G. Foerster / K. Henneberger / M. Meyer / G. F. Baryshnikov / U. Joger / W. Rosendahl / D. Nagel / R. G. Havmøller / M. T. P. Gilbert / M. Hofreiter, Historical biogeography of the leopard (*Panthera pardus*) and its extinct European populations. BMC Evolutionary Biology 18/156, 2018. DOI: 10.1186/s12862-018-1268-0.

Probst 2011: E. Probst, Eiszeitliche Leoparden in Deutschland (München 2011).

Richter/Krbetschek 2015: D. Richter / M. Krbetschek, The age of the Lower Paleolithic occupation at Schöningen. Journal of Human Evolution 89, 2015, 46-56.

Sander et al. 2006: P. M. Sander/O. Mateus/T. Laven/N. Knötschke, Bone histology indicates insular dwarfism in a new Late Jurassic sauropod dinosaur. Nature 441, 2006, 739-741.

Schüler 2010: T. Schüler, ESR-Datierung von Zahnschmelzproben der archäologischen Fundhorizonte von Neumark-Nord. In: Meller 2010a, 71-74.

Serangeli et al. 2015a: J. Serangeli / U. Böhner / T. van Kolfschoten / N. J. Conard, Overview and new results from large-scale excavations in Schöningen. Journal of Human Evolution 89, 2015, 27-45.

2015b: J. Serangeli / T. van Kolfschoten / B. Starkovitch / I. Verheijen, The European Saber-toothed cat (*Homotherium latidens*) found in the »Spear Horizon« at Schöningen (Germany). Journal of Human Evolution 89, 2015, 172-180.

Steinmetz 2005a: W.-D. Steinmetz, Das Jägerlager der Neandertaler von Lebenstedt. In: U. Joger / C. Kamcke (eds), Mammut – Elefanten der Eiszeit [exhibition catalogue] (Braunschweig 2005) 74-75.

2005b: W.-D. Steinmetz, Der Elefantenschlachtplatz von Gröbern. In: U. Joger / C. Kamcke (eds), Mammut – Elefanten der Eiszeit [exhibition catalogue] (Braunschweig 2005) 82-83.

Völker 1998: R. Völker, Hundert Jahre wissenschaftliche Grabungen in Rübeländer Höhlen. Berichte des Landesamts für Umweltschutz Sachsen-Anhalt 3, 1998, 36-38.

Zückert 1763: J. F. Zückert, Die Naturgeschichte einiger Provinzen des Unterharzes, nebst einem Anhange von den Mannsfeldischen Kupferschiefern (Berlin 1763).

SUMMARY / ZUSAMMENFASSUNG

Ice Age Fauna in the Braunschweig Region, with a Focus on the Carnivores

The Braunschweig region locates many Late Pleistocene localities (including the famous caves of the Harz Mountains and the Neanderthal site Salzgitter-Lebenstedt), but besides Schöningen, only few Middle Pleistocene sites are known in the wider area. Of these, Lehringen near Bremen, although much younger than the horse butchering site at Schöningen 13II, witnessed a comparable big game slaughter (victim: *Palaeoloxodon antiquus*). Comparable younger sites are Gröbern and Neumark-Nord in Saxony-Anhalt. Bilzingsleben in Thuringia may be slightly older than Schöningen.

As far as large carnivores are concerned, bones of cave bear *Ursus spelaeus* (in the Harz Mountains), wolf (*Canis lupus*) and lion (*Panthera leo spelaeus*) are most abundant in numerous localities. Hyena (*Crocuta crocuta spelaea*) was omnipresent too, but many records refer to biting or gnawing marks on bones of other species. Among the Upper Pleistocene localities, Hyena dens are often the origin of bone assemblages.

Among the rarer species, the leopard (*Panthera pardus*) is recorded here in some of its northernmost points of occurrence worldwide. The recent finds of *Homotherium latidens* at Schöningen add another large carnivorous species to the Pleistocene fauna of the region.

Eiszeitliche Fauna im Braunschweiger Land, mit Schwerpunkt auf den Fleischfressern

Die Braunschweiger Region beherbergt zahlreiche spätpleistozäne Fundstellen (u. a. die berühmten Höhlen des Harzes und die Neandertaler-Fundstelle Salzgitter-Lebenstedt), aber außer Schöningen sind nur wenige mittelpleistozäne Fundstellen im weiteren Umkreis bekannt. Davon ist Lehringen bei Bremen, obwohl viel jünger als der Pferdeschlachtplatz in Schöningen 13II, Zeuge einer vergleichbaren Großwildschlachtung (Beute: *Palaeoloxodon antiquus*). Vergleichbare jüngere Fundstellen sind Gröbern und Neumark-Nord in Sachsen-Anhalt. Bilzingsleben in Thüringen ist möglicherweise etwas älter als Schöningen.

Bei den Großraubtieren sind Knochen von Höhlenbär (*Ursus spelaeus*) (im Harz), Wolf (*Canis lupus*) und Löwe (*Panthera leo spelaeus*) in zahlreichen Fundorten am häufigsten vertreten. Auch die Hyäne (*Crocuta crocuta spelaea*) war allgegenwärtig, aber viele Berichte beziehen sich auf Biss- oder Nagespuren an Knochen anderer Arten. Bei den oberpleistozänen Fundorten sind Hyänenhöhlen oft der Ursprung der Knochenansammlungen.

Zu den selteneren Arten gehört der Leopard (*Panthera pardus*), der hier an einigen seiner nördlichsten Fundorte weltweit nachgewiesen wurde. Die jüngsten Funde von *Homotherium latidens* bei Schöningen ergänzen die pleistozäne Fauna der Region um eine weitere große fleischfressende Art.

ARITZA VILLALUENGA

BIOGEOGRAPHIC ANALYSIS OF UPPER PLEISTOCENE FELID (*FELIS*, *LYNX* AND *PANTHERA*) REMAINS IN ARCHAEOLOGICAL SITES ON THE IBERIAN PENINSULA

This paper will point out the geographical distribution of the felid family (*Felis, Lynx and Panthera*) bone remains in archaeological sites on the Iberian Peninsula during the Upper Pleistocene. These depositional contexts are located in caves or rock-shelters during the Upper Pleistocene, which were alternatively occupied, by both carnivores and hominins.

Before examining the presence of these species in depth on the Iberian Peninsula, some considerations must be made. Firstly, bone remains are only preserved in determinate geological contexts; and in multiple regions, karstic depositional contexts (caves, rock-shelters and potholes) have preserved bone remains. Secondly, in some regions during the 20th century the research on Palaeolithic archaeology and Quaternary palaeontology, in addition to the development and constitution of speleological groups and the exploratory work on multiple karstic systems, was encouraged. The outcome being the discovery and excavation of multiple archaeological and palaeontological remains, especially post 1970s with the appearance of specialists in zooarchaeology.

The chronological framework of this study is the Upper Pleistocene. This period begins in Western Europe with the Eemian interglacial (OIS 5e) 126 000 ± 5000 years ago and ends with the early Holocene (OIS 1), 9700 years ago. The Upper Pleistocene was dominated by the glacial phase of Würm (OIS 4, 3 and 2). Southwestern Europe was in general cold with irregular phases of warmer (interstadial) and colder (stadial) periods. During these cold periods, a continental ice sheet covered Northern Europe, and an area called the frozen desert became an unoccupied/unpopulated space. On the other hand, the southern peninsulas with milder climatic conditions were never unoccupied. The Iberian Peninsula became a climatic and geographical refuge (Sommer/Nadachowski 2006).

In this study, eight felid species have been included: the lion (*Panthera leo spelaea, Panthera leo clueti* and *Panthera* sp.), leopard (*Panthera pardus*), lynx (*Lynx lynx, Lynx pardina* and *Lynx spelaea*) and the wildcat (*Felis silvestris*). Traditionally, taxonomic classifications were done following palaeontological parameters and chrono-cultural adscription of the sedimentary contexts. Nevertheless, the identification process has evolved, and two lion forms have been identified, the cave lion (*Panthera leo spelaea*), a large carnivore often discovered in the oldest Upper Pleistocene layers, while in the younger layers a much smaller lion (*Panthera leo clueti*) appeared, and was comparable in size to the modern African lion (Castaños 2005).

Nowadays, ancient DNA analyses are providing tremendous advances in the taxonomy. During the last decade, the existence of two lion species (*Panthera leo spelaea* and *Panthera leo clueti*) has been put in doubt (Burger et al. 2004; Ersmark et al. 2014). At present, it is assumed that the Pleistocene lion decreased in size during the Upper Pleistocene, until its final extinction process at the end of the Pleistocene and eventual replacement by the European subspecies of African lion (*Panthera leo*).

The second example is the taxonomical ascription of lynx remains. Three species were identified (*Lynx lynx, Lynx spelaea* and *Lynx pardina*) on the Iberian Peninsula. In this case, the geographical location and size were fundamental to classify some of the remains as belonging to one or the other taxa. During previous

In: Nicholas J. Conard · Henning Hassmann · Kurt Felix Hillgruber · Jordi Serangeli · Thomas Terberger (eds), The *Homotherium* Finds from Schöningen 13II-4. Man and Big Cats of the Ice Age. Forschungen zur Urgeschichte aus dem Tagebau von Schöningen 4 (Mainz 2022). DOI: https://doi.org/10.11588/propylaeum.1006.c13523

decades, DNA has determined clarified taxonomic adscriptions. In 2015, the presence of the European lynx has been confirmed (*Lynx lynx*) in the Cantabrian Mountain range (northern Iberia). Furthermore, a genetic link between the Iberian lynx (*Lynx pardina*) and the Pleistocene lynx (*Lynx spelaea*) has been established (Rodriguez-Varela et al. 2015a). The Pleistocene lynx and the Iberian lynx should form a single species. The Iberian lynx evolved and adapted as a reaction to colder environments. Furthermore, the biogeographic distribution of this species spread towards the end of the Pleistocene into Southern France and Northern Italy (Rodriguez-Varela et al. 2015b).

Finally, we must address the provisionality of this paper, as in much bibliographic research, there arises the possibility that maybe some of the more obscure papers have not been cited and this is therefore incomplete. Older studies and palaeontological discoveries have not been systematically published. Decades ago, the recovery of bone remains in caves was often not even reported. These collections were kept in boxes, remaining unpublished or only mentioned in short reports. Because of this situation, it became very hard to recompile information. Fortunately, this situation has changed immensely since the 1970s, with the emergence of well-structured institutions in Spain and Portugal (departments of prehistory, natural sciences associations, museums, etc.) and the first researchers specializing in zooarchaeology slowly but surely started to achieve systems and a structured order in the conservation and recovery of finds.

Most of the felid bone remains identification was done on Pleistocene archaeological sites. During the early Upper Pleistocene, a large number of caves used as refuges and dens, with shared occupation between carnivores and hominins (Villaluenga 2016a; 2016b). Later, and in particular during the Upper Palaeolithic, the presence of smaller taxa (*Lynx* and *Felis*) on archaeological sites was linked with fur/hide exploitation or meat consumption (Yravedra 2005). The continual appearance and the absorption of new evidence of felid remains at Upper Pleistocene archaeological sites has improved the collection and compilation of information and made it state of the art.

DISCUSSION

Geography

The Iberian Peninsula is the westernmost of the three major Southern European peninsulas (Iberian, Italian and Balkan). It is bordered to the South and East by the Mediterranean Sea and on the Southwest, West and North by the Atlantic Ocean. In the Northeast the Pyrenees mountains (Aneto, 3404 m a.s.l.) separate the Iberian Peninsula from the rest of the continent, the Strait of Gibraltar separates the southern Iberian tip from Africa (Ceuta, Spain). The Iberian Peninsula extends from the southernmost extremity to the northernmost extremity over a distance of 865 km and from the easternmost to the westernmost point; the distance is approx. 1155 km. Altogether the Iberian Peninsula is surrounded by 3313 km of coastline. During the Last Glacial Maximum (LGM), this shoreline was a minimum of -115/-120 m lower than the current coastline.

The Iberian Peninsula is after the European Alps the region with the highest number of peaks. Three quarters of the Iberian Peninsula consists of a high plateau known as the »Meseta Central« ranging from 610 to 760 m in altitude. This plateau is surrounded by multiple mountain ranges where most of the main river sources are located (main rivers are Miño, Duero, Tajo, Guadiana, Guadalquivir, Segura Júcar and Ebro), which find their way to the sea through geological gaps eroded in between mountain ranges (Vera Torres 2004).

The Iberian Peninsula Mountain ranges contain rocks of multiple ages, from Ediacaran (635 Ma) to Quaternary (2.5 Ma to recent) and almost all kinds of rocks are represented, the Iberian Massif is considered to be

Fig. 1 Wildcat (*Felis silvestris*) biogeographic distribution on the Iberian Peninsula. Site numeration follows **tabs 1-4**. – (Map A. Villaluenga).

the core of the peninsula, it is composed of Mesozoic and Tertiary cover rocks. The northeastern Pyrenees and the southeast Baetic mountain range are part of the Alpine belt. Northward movement of the African tectonic plateau created these elevations. Finally, the western side formed by magmatic substrate formed by the Atlantic Ocean opening process. Erosion has produced multiple forms of evaporate and carbonate Palaeozoic, Mesozoic and Cenozoic rocks (Durán-Valsero/Robledo-Ardilla 2009). This substrate has been karstified, creating multiple forms, including caves and rock-shelters occupied by carnivores and hominins during the Pliocene and Pleistocene (Jordá-Pardo 2009).

Ethology and phylogeny

Wildcat (*Felis silvestris*)

The wildcat (*Felis silvestris*) is the smallest felid in Europe; males (7.5kg) are bigger than females (5kg). The actual distribution begins in the East in the Caucasus and continues onto the southern Iberian Peninsula in the West. Actually in the Iberian Peninsula two subspecies are identified, *Felis silvestris silvestris* in the North and

Felis silvestris tartessia in the Mediterranean and southern areas. The latter subspecies is larger and has a more contrasted coloration. This species lives in forested environments where it is able to find multiple resources and refuge. This animal has a heterogeneous diet including small mammals, birds, fish, amphibians and insects, with rabbit as its main prey on the Mediterranean coast. The wildcat has a solitary, nocturnal and territorial behaviour. This is one of the main reasons for it not to be a very well-known species (Macdonald/Barrett 2002). During the Upper Pleistocene, the wildcat is present on the whole Iberian Peninsula. It has been identified at 53 archaeological sites, from the Cantabrian Mountain range in the North to Gorham's Cave (Gibraltar) in the south including multiple sites on the Mediterranean (11) and Atlantic coasts (10) (**fig. 1; tab. 1**).

Lynx (*Lynx pardina, Lynx spelaea, Lynx lynx* and *Lynx* sp.)

There are two species of lynx in Europe: in southern Spain and Portugal the Iberian lynx, Pardel lynx or Spanish lynx (*Lynx pardinus*). This species is smaller (9-13 kg) than the European lynx (*Lynx lynx*) (males 18-30 kg and females 8-21 kg). The Iberian lynx is a specialized rabbit hunter. Its current habitat is an open forested Mediterranean environment. A high degree of specialization has put the species on the brink of extinction, with the present population fluctuating around 300 individuals, making it one of the most endangered carnivore species in the world. The European lynx (*Lynx lynx*) lives from Atlantic coast to the Taiga forests and mountains up to a height of 3400 m a.s.l. Its diet is broad, including small prey up to small and medium-sized subadult ungulates (Macdonald/Barrett 2002).

For the Upper Pleistocene a third species described in the Iberian and Southern European fossil record, the Pleistocene lynx or cave lynx (*Lynx spelaea*). The body size of this species was in between the European and the Iberian lynx. The first palaeontological description of this species made employing skeletal remains discovered at Grimaldi (Boule 1919) and L'Observatoire (Boule/de Villeneuve 1927) caves (Monaco). However, only in the 1980s was it officially accepted as being a distinct species. Finally, DNA analysis has proved a direct genetic link between the Iberian lynx (*Lynx pardinus*) and the Pleistocene lynx (*Lynx spelaea*) (Rodriguez-Varela et al. 2015a). Following the most recent information it considered as a single species, nevertheless for this paper we have kept the distinct nomenclature for both forms (**fig. 2**). The Iberian lynx distribution was wide, including the Iberian Peninsula, the French Mediterranean coast and Northern Italy during the late Upper Pleistocene and early Holocene.

European lynx (*Lynx lynx*) was identified for the first time during the 1980s in the southern Pyrenees at the Upper Palaeolithic archaeological sites Rascaño (prov. Cantabria/E; González-Echegaray/Barandiarán 1981) and Erralla (prov. Gipuzkoa/E; Altuna/Baldeon/Mariezkurrena 1985). Later on, more remains were identified on the Mediterranean coast. The Ebro River Valley acts as a natural boundary for this species (**fig. 3**).

Leopard (*Panthera pardus*)

The presence of leopard (*Panthera pardus*) remains on the Iberian Peninsula begins during the Middle Pleistocene and appeared regularly in archaeo-palaeontological contexts until the end of the Upper Pleistocene (**fig. 4**). The extinction of the Pleistocene leopard came during the Magdalenian (climatic phase Würm IV) and early Holocene (Castaños 1987; Sanchís et al. 2015) as the most probable causes of extinction mentioned are the climatic changes and the changes in the leopard prey population.

To try to have an idea of what this extinct leopard was like, we must look at the actual single leopard living presently in the northern environments, the Amur leopard (*Panthera pardus orientalis*). This species lives

Fig. 2 Iberian lynx (*Lynx pardinus,* in yellow) and Pleistocene or cave lynx (*Lynx spelaea,* in red) biogeographic distribution on the Iberian Peninsula. Site numeration follows **tabs 1-4**. – (Map A. Villaluenga).

in Eastern Siberia. Its body size should be similar to the Pleistocene European leopard (31-70kg), with a length up to 190cm and a height of 80cm. Quite exceptionally this subspecies has two furs, one for winter (7cm) and one for summer (2cm) the last one being more colourful. Leopards have preferential nocturnal behaviour; during the day they usually rest in rock-shelters, caves or on tree branches.

The leopard is a solitary ambush hunter. Ethological sources have documented that sometimes they would bring their prey carcasses to caves and rock-shelters to protect them from scavengers, these refuges are also used as dens (Ruiter/Berger 2000). This behaviour should be present in European Pleistocene leopards, some Upper Pleistocene archaeozoological assemblages have been interpreted as leopard consumption sites, characterized by middle size ungulate remains, in particular ibex and chamois (Yravedra 2007a).

Lion (*Panthera spelaea* and *Panthera leo clueti*)

The cave lion is one of the most popular species of the Eurasian Prehistory. During decades of work on the fossil record, two subspecies of lion have been identified: *Panthera leo clueti* with a body size similar to modern-day lions (>200kg) and the larger *Panthera spelaea goldfuss* (Ballesio 1980). Actually, this dis-

Fig. 3 European lynx (*Lynx lynx* in yellow) and indeterminate lynx (*Lynx* sp.) biogeographic distribution on the Iberian Peninsula. Site numeration follows **tabs 1-4**. – (Map A. Villaluenga).

tinction attributed to a size reduction through the Upper Pleistocene: older animals were larger than the younger individuals were, these should be comparable to actual African lions (Burger et al. 2004). Some other authors consider that these size differences are due to sexual dimorphism (Turner 1984), as extant male lions are 30 % larger than females.

For decades the origins of the large felids was not clear. Nowadays, most of the researchers consider that the most probable hypothesis is that 4.5 million years ago *Panthera gombaszoegensis* was the ancestor of the lion, tiger, leopard and jaguar. Lion (*Panthera leo*) and cave lion (*Panthera spelaea*) evolved from the first lion species (*Panthera leo fossilis*), appearing around 3 million years ago in Eastern Africa. The direct ancestor of the cave lion migrated to Eurasia 900 000 years ago. Since then, the European and African lions have evolved independently in four populations: the African lion (*Panthera leo*), the cave lion (*Panthera spelaea*) the Asiatic lion (*Panthera leo persica*) and the American lion (*Panthera leo youngi*). Last one migrated to North America during the Middle Pleistocene (400 000-300 000 years ago) and evolved in the Upper Pleistocene *Panthera leo atrox* (Barnett et al. 2009; Ersmark et al. 2014). During the early Holocene in Southern Europe, after the extinction (10 000 years ago) of the cave lion (*Panthera spelaea*) (Stuart/Lister 2011), Europe was colonized by another lion subspecies (*Panthera leo europaea*), who maybe originated from African or Persian lion population. Maybe this subspecies has been identified

Fig. 4 Leopard (*Panthera pardus*) biogeographic distribution on the Iberian Peninsula. Site numeration follows **tabs 1-4**. – (Map A. Villaluenga).

as *Panthera leo clueti* in southwestern European late Pleistocene and early Holocene archaeological and palaeontological contexts.

Further analysis is necessary in this aspect. This last European lion (*Panthera leo europaea*) colonized Southwestern Europe, from the Iberian Peninsula (**fig. 5**) to the Balkans, being a popular prey amongst the Greeks and Romans, the over-hunting of this animal (100 AD) led to its extinction on the Balkans.

The cave lion's (*Panthera spelaea*) external appearance is still uncertain; Upper Palaeolithic art has been for a while the single source of information. Chauvet Cave (dép. Ardèche/F) contains a whole set of lion representations, but not including the male lions characteristically mane. This is the case in some other caves as Combarelles (dép. Dordogne/F), Armintxe (Lekeitio, Biscay/E) or Vogelherd (Swabian Jura/D) among others. However, the absence of this feature is usual in the Indian lion (Azemá 2009; 2010). The ethology of this species is unknown, single living examples are Indian (*Panthera leo persica*) and African lions (*Panthera leo*), both live in groups, composed of females, cubs and one or two dominant males. They defend a well-defined territory, located in open forested spaces. The diet of lions is composed of middle-sized ungulates and in Africa occasionally by megafauna. Scavenging is also a recurrent behaviour. The cave lion (*Panthera spelaea*) was a powerful hunter, probably in direct competition with the cave hyena (*Crocuta crocuta spelaea*) and hominins. Isotopic analyses has revealed a diet composed of horse, bison, aurochs, red deer

Fig. 5 Biogeographic distribution of lion on the Iberian Peninsula. European late Pleistocene and early Holocene lion (*Panthera leo clueti* in yellow), cave lion (*Panthera spelaea* in red) and indeterminate lion (*Panthera* sp. in blue). Site numeration follows **tabs 1-4**. – (Map A. Villaluenga).

and Pleistocene megafauna, as the giant-deer or subadult woolly rhinoceros and mammoth (Bocherens et al. 2011).

Spatial distribution

Most archaeological sites on the Iberian Peninsula are located within karstic contexts. This geological context is chemically dissolved by water creating multiple forms and orifices as caves and rock-shelters. These structures inhabited during the Pleistocene by hominins and carnivores, the following lines will present the location of felid remains across the Iberian Peninsula.

As mentioned, remains preservation is linked to multiple factors: First, the geology and the distribution of appropriate substrates. Second, the biogeographic distribution of the species. For example, some species populated the entire territory, while the European lynx (*Lynx lynx*) and cave lion (*Panthera spelaea*) were restricted to the northern and central areas. Third, the archaeological field work tradition, the influence of the first French researchers on prehistoric archaeology as they introduced this discipline at the end of the

19th century and the early 20th century. Fourth, the preservation of bone remains. Organic remains on the Iberian Peninsula are only preserved in exceptional circumstances, fluvial-lacustrine and karstic contexts (Sala-Ramos et al. 2014). Considering these criteria, it should be possible to correct the overrepresentation of northern Iberia sites comparing to the rest of the peninsula.

The spatial distribution of felids (*Felis silvestris*, *Lynx lynx*, *Lynx pardinus*, *Lynx spelaea, Panthera pardus, Panthera leo, Panthera spelaea* and *Panthera* sp.) shows differences. Some taxa like the wildcat (*Felis silvestris*), the Iberian lynx (*Lynx pardinus*) and the leopard (*Panthera pardus*) are present over the whole peninsula. On the other hand, the distribution of the European lynx (*Lynx lynx*) is restricted to the northeastern Cantabrian region. Lions (*Panthera spelaea* and *Panthera leo*) are restricted to the northern half of the Iberian Peninsula. A limited distribution to Northern Iberia is not exclusive to these species: the distribution of reindeer, woolly rhinoceros, mammoth (García/Arsuaga 2003), cave bear (Villaluenga 2009), arctic fox (Altuna/Mariezkurrena 2004) and wolverine (Altuna 1963; Altuna/Baldeón 1986) is limited to the northern part of the peninsula. This model of distribution could be related to climatic and environmental conditions, linking the northern part of the Iberian Peninsula to the Upper Pleistocene and glacial, non-analogue continental environments. In contrast, warmer conditions should be present in the Mediterranean area and the southern part of the peninsula.

In total, 143 archaeological sites have been analysed. To understand the different regional patterns, the Iberian Peninsula have been divided into five different regions: Cantabrian Corniche, Centre and Pyrenees, Mediterranean coast, South and West. The Cantabrian Corniche with a geological substrate composed of carbonate massifs has the highest number of sites (57) and quantitatively the highest number of bone remains and species. Furthermore, a long tradition of Quaternary palaeontology and prehistoric archaeology has helped to encourage multiple excavation projects and zooarchaeological analysis, including regional biogeographic and chronologic analysis (Altuna 1992a; 1992b; Castaños 1987; 2005) (see **tab. 1**). The Central Iberian Peninsula and Pyrenees were peopled since the early Upper Pleistocene, however, the Pleistocene deposits containing organic remains are few (18) because the preservation in this region only occurred in rock-shelters or caves (**tab. 2**; Blasco-Sancho/Montes/Utrilla 1996; Yravedra 2007b; Sauqué et al. 2014). The Mediterranean coast is revealing more and more archaeological accumulations (40), due to a systematic analysis by multiple research teams (**tab. 3**; Estévez 1979; Sanchís 2015; Sanchís et al. 2015). The southern region is considered an exceptional territory with very diverse environments, from the dry Mediterranean and Atlantic coastal lines to high altitude occupations in the Baetic Massif (**tab. 4**; Barroso et al. 2014; Finlayson et al. 2014). In this region, only ten excavated sites show the presence of bone remains covering four separate felid species. Finally, the western region (17) has an intense site concentration around the Tajo River mouth (**tab. 4**; Cardoso 1993). On the Iberian Peninsula the wildcat (*Felis silvestris*) has been identified at 53 sites, the lynx (*Lynx lynx, Lynx spelaea* and *Lynx pardinus*) has been found at 80 sites, the leopard (*Panthera pardus*) at 76 sites and the lion (*Panthera spelaea, Panthera leo* and *Panthera leo* sp.) at 31 sites.

An individualized comment is required for each species. Firstly, *Felis silvestris* is present over the whole Peninsula, being represented at 53 sites (**fig. 1**). However, remains are more frequent in the Cantabrian Corniche (21) when compared with the remaining regions, but considering the percentage as a reference, the western region is the area with a higher (58 %) concentration of wildcat in its faunal assemblage.

Lynx (*Lynx pardinus, Lynx spelaea, Lynx lynx* and *Lynx* sp.) split into two figures: includes the Iberian lynx (*Lynx pardinus*) and the Pleistocene lynx (*Lynx spelaea*) (**figs 2-3**). As previously mentioned, both are currently considered two morphotypes of a single species (Rodriguez-Varela et al. 2015a). The presence of the Iberian lynx recorded at 57 sites; the highest frequency documented in the northeastern region (16 sites) and the northern Mediterranean coast (14 sites). The Iberian lynx (*Lynx pardinus* and *Lynx spelaea*) is present over three quarters of the peninsula, its absence from the northwestern area could be biogeographically and palaeoenvironmentally significant.

European lynx (*Lynx lynx*) and indeterminate lynx (*Lynx* sp.) remains are restricted to the eastern half of the Iberian Peninsula (**fig. 3**). First taxa has a northern distribution (Cantabrian Corniche and Mediterranean coast) being present on the Iberian Peninsula during cold climatic phases (García/Arsuaga 2003). The European lynx colonized the Cantabrian region with a higher density in the eastern area (8) than in the central (5) and western (1) areas. On the eastern extreme of the Iberian Peninsula, seven sites have reported presence of this species, and actual data shows that the Ebro River acts as a natural boundary to the spatial distribution of the European lynx on the Iberian Peninsula. However, a single exception must be reported in Cova del Corb (Ondara, prov. Alicante/E). This taxonomic determination should be confirmed. Indeterminate lynx remains published from 20 archaeological sites, most of them on the Mediterranean coast (10), the southern extreme (4), the central mountain range (2) and finally three sites in the Cantabrian Corniche. To conclude with this general distribution, a total absence of lynx bone remains in archaeological sites on the northwestern peninsula must be mentioned, an area with few archaeological records but including leopard (*Panthera pardus*) and lion (*Panthera leo* sp.) remains.

Large felids are also present on the Iberian Peninsula: the leopard (*Panthera pardus*) was the carnivore with the highest number of recovered remains and was present on 76 archaeological sites (**fig. 4**). In two of the regions, the leopard identified in one-third of all archaeological sites: 26 sites (36 %) in the Cantabrian Corniche and 22 in the Mediterranean region (35 %). On the other hand, in the centre of the Iberian Peninsula and the Pyrenees, the leopard (*Panthera pardus*) has been identified at eleven sites (61 %), followed by seven sites in the southern area (70 %) and western region (58 %). As mentioned before, maybe the repetitive presence in karstic contexts should be related to the ethology, as the leopard probably was a solitary ambusher, specializing in medium and small size ungulates. Bone remains, gnawed by leopards were identified as ibex (*Capra pyrenaica*) and chamois (*Rupicapra pyrenaica*) (Yravedra 2007a), typical species for rocky biotopes.

Lion (*Panthera leo clueti, Panthera spelaea* and *Panthera* sp.) is the largest and least represented carnivore in the Iberian Peninsula (36 sites) (**fig. 5**). *Panthera leo clueti* (9) and *Panthera spelaea* (7) have been identified in the Cantabrian Corniche (Castaños 2005). Taxonomic differentiation based on body size, as *Panthera spelaea* was larger than any actual felid. For a while, *Panthera leo clueti* has been considered as a modern species incoming during the late Upper Pleistocene, which survived until the early Holocene. However, DNA analysis has revealed a genetic relation between both Upper Pleistocene lions. Nowadays, most of the researchers agree on the existence of a gradual size diminution through time. If this is correct, *Panthera spelaea* and *Panthera leo clueti* should be considered as a single taxa, known as *Panthera leo spelaea* (P. Fosse, personal communication). Later, after the extinction of this species (Stuart/Lister 2011) a modern lion, the European subspecies (*Panthera leo europaea*) colonized Southern Europe up until the early Holocene.

Over the rest of the Iberian Peninsula, the lion has been identified as *Panthera* sp. or *Panthera leo* sp. in 20 archaeological sites, apparently this taxa was missing in Southern Iberia, the southernmost distribution reached the Tajo River mouth (Gruta do Escoural; prov. Alentejo/E) in the West and Cap del Nau (Cova Negra; prov. Alicante/E) in the East. Nevertheless, in Central Iberia, the Central Massif (Cueva de los Casares and Jarama VI [both prov. Guadalajara/E]) is considered the natural boundary of the biogeographic distribution of *Panthera leo* sp. or *Panthera* sp.

CONCLUSIONS

In this paper I have analysed felid (*Felis silvestris, Lynx pardinus, Lynx lynx, Lynx* sp., *Panthera pardus, Panthera leo clueti, Panthera spelaea* and *Panthera* sp.) presence in 143 archaeological sites on the Iberian Peninsula during the Upper Pleistocene. This study provides a preliminary view on felid biogeographic distribution.

Overall, it is possible to identify a heterogeneous site distribution, linked to the presence of the appropriate geological substrate to provide refuges, rock-shelters and caves. These areas are mainly located in karstic areas near coastlines. Differences in felid presence is also related to the ecological and climatic preferences of each species. The Iberian lynx (*Lynx pardinus*) is a specialized rabbit and small prey hunter. Its population and eventual presence in a region related to the distribution of prey. However, the Pleistocene lynx (*Lynx spelaea*) distribution was restricted to the northern half of the peninsula. This difference in biogeographic distribution could be related to recent (end of the Upper Pleistocene) adaptations, specialization and body size reduction to warmer and dryer environmental conditions.

Also taxa not represented on the full territory, such as the European lynx (*Lynx lynx*) and lion (*Panthera leo clueti, Panthera [leo] spelaea* and *Panthera* sp.), should be related to multiple degrees of adaptation to/in colder environments. The distribution area of the European lynx just reaches the Cantabrian Corniche and the Northern Mediterranean coast, while lion remains extend to the southern third of the Iberian Peninsula.

Finally, biogeographic analysis should become a powerful tool used to make interpretations in prehistoric and quaternary environmental analysis. However, these studies are in a state of constant evolution and this paper has been focused on a single family. In the future, this type of study should be extended to include large and small mammals. Furthermore, new analyses will provide data able to answer certain questions, such as the reasons for the absence of certain taxa in some areas of the region. The Iberian Peninsula, a relatively restricted geographical region revealed a complex territory, with multiple environments and geographic conditions. These factors have provided the geographic overlap of species adapted to various different conditions making the Iberian Peninsula a unique region for the analysis of faunal adaptations to climatic changes.

ACKNOWLEDGEMENTS

This research was possible thanks to Basque Government PhD Research Grant Program funding, High yield research group on Prehistory (IT-1223-19) University of the Basque Country (Vitoria-Gasteiz, Spain) (aritza.villaluenga@ehu.eus), Aranzadi Sciences Society (Saint Sebastian, Spain) and MONREPOS Research Centre and Museum for Human Behavioural Evolution-RGZM, the support. – Finally, I would like to acknowledge Felix Hillgruber and PALÄON team for the invitation to the *Homotherium* workshop.

REFERENCES

Altuna 1963: J. Altuna, Primer hallazgo de glotón (*Gulo gulo* L.) en la Península Ibérica. Munibe 15, 1963, 128.

1992a: J. Altuna, Asociaciones de macromamíferos del Pleistoceno Superior en el Pirineo Occidental y Cantábrico. In: A. Caerreta / F. M. Ugarte (eds), The Late Quaternary in the Western Pyrenean Region. Proceedings of the International Conference on the Environment and the Human Society in the Western Pyrenees and the Basque Mountains during the Upper Pleistocene and the Holocene; Gasteiz/Vitoria, May 3-5, 1990 (Bilbao 1992).

1992b: J. Altuna, El medio ambiente durante el Pleistoceno Superior en la región cantábrica con referencia especial a sus faunas de mamíferos. Munibe [Antropologia, Arkeologia] 44, 1992, 13-29.

Altuna/Baldeón 1986: J. Altuna / A. Baldeón, Resultados del sondeo estratigráfico practicado en la cueva de Mairulegorreta XI, Gorbea (Álava). Estudios de Arqueología Alavesa 13, 1986, 47-62.

Altuna/Mariezkurrena 2004: J. Altuna / K. Mariezkurrena, Los hallazgos de zorros (*Vulpes vulpes* Linnaeus y *Alopex lagopus* Linnaeus) en yacimientos paleolíticos de la región cantábrica. In: E. Baquedano Pérez / S. Rubio Jara (eds), Miscelánea en homenaje a Emiliano Aguirre. 2: Paleontología. Zona Arqueológica 4 (Madrid 2004) 50-54.

Altuna/Baldeon/Mariezkurrena 1985: J. Altuna / A. Baldeon / K. Mariezkurrena (eds), Cazadores magdalenienses en Erralla (Cestona, País Vasco). Munibe [Antropologia, Arkeologia] 37 (San Sebastián 1985).

Azemá 2009: M. Azemá, L'art des cavernes en action. 1: Les animaux modèles: aspect, locomotion, comportement (Paris 2009).

2010: M. Azemá, L'art des cavernes en action. 2: Les animaux figurés: animation et mouvement, l'illusion de la vie (Paris 2010).

Ballessio 1980: R. Ballesio, Le gisement pléistocène supérieur de la grotte de Jaurens à Nespouls, Corrèze, France: les Carnivores (Mammalia, Carnivora). II. Felidae. Nouvelles Archives du Muséum d'Histoire Naturelle de Lyon 13, 1980, 68-97.

Barnett et al. 2009: R. Barnett / B. Shapiro / I. Barnes / S. Y. W. Ho / J. Burger / N. Yamaguchi / T. F. G. Higham / H. T. Wheeler / W. Rosendahl / L. V. Sher / M. Sotnikova / T. Kuznetsova / G. F. Baryshnikov / L. D. Martin / R. Harington / J. A. Burns / A. Cooper, Phylogeography of Lions (*Panthera leo* ssp.) Reveals Three Distinct Taxa and a Late Pleistocene Reduction in Genetic Diversity. Molecular Ecology 18/8, 2009, 1668-1677.

Barroso et al. 2014: C. Barosso / M. Caparrós / D. Barsky / A. M. Moigne / A. Monclova, Cueva del Boquete de Zafarraya: un yacimiento de neandertales en el sur Iberia. In: Sala Ramos et al. 2014, 463-472.

Blasco-Sancho/Montes/Utrilla 1996: M. F. Blasco-Sancho / L. Montes / P. Utrilla, Deux modeles de strategie occupationelle dans le Mousterien tardif de la Vallee de l'Ebre: les grotes de Peña Miel et Gabasa. In: E. Carbonell / M. Vaquero (eds), The Last Neanderthals, the First Anatomically Modern Humans. A Tale about the Human Diversity. Cultural Change and Human Evolution. The Crisis at 40 ka BP (Tarragona 1996) 289-313.

Bocherens et al. 2011: H. Bocherens / D. G. Drucker / D. Bonjean / A. Bridault / N. J. Conard / C. Cupillard / M. Germonpré / M. Höneisen / S. C. Münzel / A. Napierala / M. Patou-Mathis / E. Stephan / H.-P. Uerpmann / R. Ziegler, Isotopic Evidence for Dietary Ecology of Cave Lion (*Panthera spelaea*) in North-Western Europe: Prey Choice, Competition and Implications for Extinction. Quaternary International 245/2, 2011, 249-261.

Boule 1919: M. Boule, Les grottes de Grimaldi (Baoussé-Roussé). I, 4: Géologie et paléontologie (Monte Carlo 1919).

Boule/de Villeneuve 1927: M. Boule / L. de Villeneuve, La grotte de L'Observatoire à Monaco. Archives de l'Institut de Paléontologie Humaine, Mémoire 1 (Paris 1927).

Burger et al. 2004: J. Burger / W. Rosendahl / O. Loreille / H. Hemmer / T. Eriksson / A. Götherström / J. Hiller / M. J. Collins / T. Wess / K. W. Alt, Molecular Phylogeny of the Extinct Cave Lion *Panthera leo spelaea*. Molecular Phylogenetics and Evolution 30, 2004, 841-849.

Cardoso 1993: J. L. Cardoso, Contribuçao para o Conhecimento dos grandes Mamíferos do Pleistoceno Superior de Portugal (Oeiras 1993).

Castaños 1987: P. Castaños, Los carnívoros prehistóricos de Vizcaya. Kobie [Serie Paleoantropología] 16, 1987, 7-50.

2005: P. Castaños, Estudio paleontológico de un esqueleto de león, *Panthera leo* de la Sima de Azoleta, Gorbeia, Alava. Munibe [Antropologia, Arkeologia] 57, 2005, 123-129.

Dúran-Valsero/Robledo-Ardila 2009: J. J. Durán-Valsero / P. A. Robledo-Ardila, Carbonate and Evaporite karst systems of the Iberian Peninsula and the Balearic Islands. In: A. García-Cortés / J. Águeda-Villar / J. Palacio Suárez-Valgrande / C. I. Salvador González (eds), Spanish Geological Frameworks and Geosites, an Approach to Spanish Geological Heritage of International Relevance (Madrid 2009) 200-215.

Ersmark et al. 2014: E. Ersmark / L. Orlando / E. Sandoval-Castellanos / I. Barnes / R. Barnett / A. Stuart / A. Lister / L. Dálen, Population Demography and Genetic Diversity in the Pleistocene Cave Lion. Open Quaternary1, 2014, p.Art. 4. DOI: 10.5334/oq.aa.

Estévez 1979: J. Estévez, La Fauna del Pleistoceno catalán [Diss. Univ. Autònoma de Barcelona 1979].

Finlayson et al. 2014: C. Finlayson / R. Blasco / J. Rodriguez-Vidal / F. Giles-Pacheco / G. Finlayson / J. M. Gutierrez / R. Jennings / D. A. Fa / J. Rosell / J. M. Carrión / A. Sanchez-Marco / S. Finlayson / M. A. Bernal, Excavaciones en Gibraltar: especial referencia a Gorham's cave y Vanguard Cave. In: Sala Ramos et al. 2014, 506-514.

García/Arsuaga 2003: N. García / J. L. Arsuaga, Last Glaciation cold-adapted faunas in the Iberian Peninsula. In: J. W. F. Reumer / J. de Vos / D. Mol (eds), Advances in Mammoth Research. Proceedings of the Second International Mammoth Conference, Rotterdam, 16-20 July 1999). Deinsea 9 (Rotterdam 2003) 159-169.

González Echegaray/Barandiarán 1981: J. González Echegaray / I. Barandiarán, El Paleolítico Superior de la Cueva del Rascaño, Santander (Santillana del Mar, Santander 1981).

Jordá-Pardo 2009: J. F. Jordá-Pardo, Vertebrate deposits of the Spanish Pliocene and Pleistocene. In: A. García-Cortés / J. Águeda-Villar / J. Palacio Suárez-Valgrande / C. I. Salvador González (eds), Spanish Geological Frameworks and Geosites, an Approach to Spanish Geological Heritage of International Relevance (Madrid 2009) 171-183.

Macdonald/Barrett 2002: D. W. Macdonald / P. Barrett, Mammals of Europe (Princeton NJ 2002).

Rodriguez-Varela et al. 2015a: R. Rodriguez-Varela / N. García / C. Nores / D. Álvarez-Lao / R. Barnett / J. L. Arsuaga / C. Valdiosera, Ancient DNA Reveals Past Existence of Eurasian Lynx in Spain. Journal of Zoology 298/22, 2015. DOI: 10.1111/jzo.12289.

2015b: R. Rodriguez-Varela / A. Tagliacozzo / I. Ureña / N. García / E. Crégut-Bonnoure / M. A. Mannino / J. L. Arsuaga / C. Valdiosera, Ancient DNA Evidence of Iberian Lynx Palaeoendemism. Quaternary Science Reviews 112, 2015, 172-180.

Ruiter/Berger 2000: J. D. Ruiter / L. R. Berger, Leopard as Taphonomic Agents in Dolomitic Caves. Implications for Bone Accumulations in the Hominid Bearning Deposits of South Africa. Journal of Archeological Science 27, 2000, 665-684.

Sala Ramos et al. 2014: R. Sala Ramos / E. Carbonell / J. M. Bermúdez de Castro / J. L. Arsuaga (eds), Los cazadores recolectores del Pleistoceno y del Holoceno en Iberia y el Estrecho de Gibraltar, estado actual del conocimiento del registro arqueológico (Burgos 2014).

Sanchís-Serra 2015: A. Sanchís-Serra, Un món de feres. Grans carnívors en la prehistòria valenciana. Museu de Prehistòria de València (Valencia 2015).

Sanchís et al. 2015: A. Sanchís / C. Tormo / V. Sauqué / V. Sanchís / R. Díaz / A. Ribera / V. Villalverde, Pleistocene Leopards in the Iberian Peninsula: New Evidence from Palaeontological and Archaeological Contexts in the Mediterranean Region. Quaternary Science Reviews 124, 2015, 175-208.

Sauqué et al. 2014: V. Sauqué / R. Rabal-Garcés / C. Sola-Almagro / G. Cuenca-Bescós, Bone Accumulation by Leopards in the Late Pleistocene in the Moncayo Massif (Zaragoza, NE Spain). PLoS ONE 9/3, 2014, e92144. DOI: 10.1371/journal.pone.0092144.

Sommer/Nadachowski 2006: R. S. Sommer / A. Nadachowski, Glacial Refugia of Mammals in Europe: Evidence from Fossil Records. Mammal Review 36/4, 2006, 251-265.

Stuart/Lister 2011: J. A. Stuart / A. M. Lister, Extinction chronology of the cave lion *Panthera spelaea*. Quaternary Science Reviews 30/17-18, 2011, 2329-2340.

Turner 1984: A. Turner, Dental Sex Dimorphism in European Lions (*Panthera leo* L.) of the Upper Pleistocene: Palaeoecological and Palaeoethological Implications. Annales Zoologici Fennici 21/1, 1984, 1-8.

Vera Torres 2004: J. A. Vera Torres, Geología de España (Madrid 2004).

Villaluenga 2009: A. Villaluenga, Yacimientos del Pleistoceno Superior en la Península Ibérica con presencia de restos de oso. Munibe [Antropologia, Arkeologia] 60, 2009, 17-33.

2016a: A. Villaluenga, Úrsidos en medios kársticos de la Cornisa Cantábrica (País Vasco y Navarra). Estudio tafonómico de conjuntos arqueológicos y paleontológicos del Pleistoceno Superior y Holoceno. BAR International Series 2810 (Oxford 2016).

2016b: A. Villaluenga, Presencia de felinos (*Felis*, *Lynx* y *Panthera*) en el registro arqueológico de la Península Ibérica durante el Pleistoceno Superior. Archaeofauna 25, 2016, 171-190.

Yravedra 2005: J. Yravedra, Aprovechamiento càrnico de lince (*Lynx pardina*) durante el Pleistoceno Superior en el interior de la Península Ibérica. Munibe [Antropologia, Arkeologia] 57, 2005, 303-311.

2007a: J. Yravedra, Nuevas contribuciones en el comportamiento cinegético de la cueva de Amalda. Munibe [Antropologia, Arkeologia] 58, 2007, 43-88.

2007b: J. Yravedra, Aproximaciones tafonómicas a los cazadores de la segunda mitad del pleistoceno superior de la mitad norte del interior de la Península Ibérica. Arqueoweb. Revista Sobre Arqueología en Internet 9/1, 2007. www.ucm.es/info/arqueoweb/pdf/9-2/josete.pdf (10.9.2020).

SUMMARY / ZUSAMMENFASSUNG

Biogeographic Analysis of Upper Pleistocene Felid (*Felis*, *Lynx* and *Panthera*) Remains in Archaeological Sites on the Iberian Peninsula

This paper aims to introduce the biogeographic distribution of felids (*Felis, Lynx* and *Panthera*). The analysed period is within the last glacial phases. This period was quite unstable, changing climatic conditions affected mammals biogeographic distribution. I would like to focus this analysis on the Iberian Peninsula, the southwestern most region of Europe. However, it is a complex geography with different climatic areas affect/ing the species distribution.
Eight felid species have been identified on the Iberian Peninsula, from small carnivores such as the wildcat (*Felis silvestris*), up to the middle-sized lynx (*Lynx pardinus, Lynx lynx* and *Lynx spelaea*), and big carnivores such as the leopard (*Panthera pardus*) and several species of lion (*Panthera leo clueti, Panthera spelaea* and *Panthera* sp.). A diachronic and biogeographic analysis of this carnivore family allows a better comprehension of mammal adaptations to changing environments.

Biogeographische Analyse von oberpleistozänen Feliden (*Felis*, *Lynx* und *Panthera*) Überresten in archäologischen Stätten auf der Iberischen Halbinsel

Dieses Papier gibt eine Einführung in die biogeographische Verteilung von Feliden (*Felis*, Luchs und *Panthera*). Der analysierte Zeitraum liegt innerhalb der letzten Vereisungsphasen. Dieser Zeitraum war recht instabil, da sich die wechselnden klimatischen Bedingungen auf die biogeographische Verteilung der Säugetiere auswirkten. Ich konzentriere mich bei dieser Analyse auf die Iberische Halbinsel, die südwestlichste Region Europas. Es handelt sich jedoch um eine komplexe Geographie mit unterschiedlichen klimatischen Gebieten, die sich auf die Verbreitung der Arten auswirken.
Auf der Iberischen Halbinsel wurden acht Felidarten identifiziert, von kleinen Fleischfressern wie der Wildkatze (*Felis silvestris*) bis hin zum mittelgroßen Luchs (*Lynx pardinus*, *Lynx lynx* und *Lynx spelaea*) und großen Fleischfressern wie dem Leoparden (*Panthera pardus*) und mehrere Löwenarten (*Panthera leo clueti*, *Panthera spelaea* und *Panthera* sp.). Eine diachrone und biogeographische Analyse dieser Fleischfresserfamilie ermöglicht ein besseres Verständnis der Anpassungen der Säugetiere an Umweltveränderungen.

no.	site	location	*Felis silvestris*	*Lynx pardina*	*Lynx spelaea*	*Lynx lynx*	*Lynx* sp.	*Panthera pardus*	*Panthera leo*	*Panthera spelaea*	*Panthera leo* sp.
1	Cueva Eirós	Triacastela, Lugo						×			
2	Las Caldas	Priorio, Asturias						×			
3	La Güelga	Cangas de Onís, Asturias						×			
4	La Riera	Posada de Llanes, Asturias						×	×	×	
5	Llonín	Peñamera Alta, Asturias						×			
6	La Paloma	Soto de Regueras, Asturias	×						×		
7	Colombres	Ribadeva, Asturias	×								
8	Cueto de la Mina	Llanes, Asturias	×								
9	Tudela Veguín	Grado, Asturias	×								
10	Sima del Sueve	El Sueve, Asturias				×					
11	Tito Bustillo	Ribadesella, Asturias			×						
12	Balmorí	Llanes, Asturias									×
13	Quintanal	Llanes, Asturias									×
14	La Parte	Siero, Asturias								×	
15	Hornos de la Peña	S. F. de Buelna, Cantabria				×		×			
16	Castillo	Puente Viesgo, Cantabria						×	×	×	
17	El Juyo	Igollo, Cantabria						×			×
18	Morín	Villaescusa, Cantabria	×		×	×		×			
19	El Mirón	Ramales, Cantabria	×								×
20	El Otero	Voto, Cantabria	×								
21	Covalejos	Arce, Cantabria	×								
22	Valle	Ramales, Cantabria	×								

Tab. 1 Archaeological sites at the Cantabrian Corniche, location of the sites and felid species presence.

no.	site	location	*Felis silvestris*	*Lynx pardina*	*Lynx spelaea*	*Lynx lynx*	*Lynx* sp.	*Panthera pardus*	*Panthera leo*	*Panthera spelaea*	*Panthera leo* sp.
23	Rascaño	Miera, Cantabria				×					
24	Altamira	Santillana del Mar, Cantabria				×	×		×		
25	Cueva San Juan	Arredondo, Cantabria						×			
26	El Pendo	Camargo, Cantabria							×		×
27	Covacho Arenillas	Islares, Cantabria						×			
28	El Cuco	Castro Urdiales, Cantabria						×			
29	Las Pajucas	Lanestosa, Bizkaia				×		×			
30	Arlanpe	Lemoa, Bizkaia						×			×
31	Atxuri	Mañaria, Bizkaia						×			
32	Oyalkoba	Abadiano, Bizkaia						×			
33	Bolinkoba	Abadiano, Bizkaia		×				×			
34	Axlor	Dima, Bizkaia				×		×			
35	Goikolau	Mañaria, Bizkaia	×								
36	Lumentxa	Lekeitio, Bizkaia	×			×	×				
37	Santimamiñe	Kortezubi, Bizkaia	×			×			×		
38	Lamiñak II	Berriatua, Bizkaia	×								
39	Urratxa III	Gorbea, Bizkaia				×					
40	Venta Laperra	Carranza, Bizkaia		×							
41	El Polvorín	Carranza, Bizkaia		×							
42	Arrillor	Zigoitia, Araba	×	×		×					
43	Peñas de Oro	Zuia, Araba	×								
44	Praileaitz I	Deba, Gipuzkoa						×			
45	Astigarraga	Deba, Gipuzkoa	×			×		×			
46	Ekain	Deba, Gipuzkoa						×			

Tab. 1 (continued)

no.	site	location	*Felis silvestris*	*Lynx pardina*	*Lynx spelaea*	*Lynx lynx*	*Lynx* sp.	*Panthera pardus*	*Panthera leo*	*Panthera spelaea*	*Panthera leo* sp.
47	Amalda	Zestoa, Gipuzkoa		×	×			×	×	×	
48	Lezetxiki	Arrasate, Gipuzkoa	×	×		×		×		×	
49	Labeko Koba	Arrasate, Gipuzkoa	×								
50	Aitzbitarte III	Errenteria, Gipuzkoa						×			
51	Marizulo	Urnieta, Gipuzkoa	×								
52	Urtiaga	Deba, Gipuzkoa	×	×		×			×	×	
53	Ermittia	Deba, Gipuzkoa		×							
54	Aitzbitarte IV	Errenteria, Gipuzkoa								×	
55	Abauntz	Arraitz, Navarra	×					×	×		
56	Coscobilo	Olazagutia, Navarra						×			
57	Zatoya	Abaurrea Alta, Navarra		×							

Tab. 1 (continued)

no.	site	location	*Felis silvestris*	*Lynx pardina*	*Lynx spelaea*	*Lynx lynx*	*Lynx* sp.	*Panthera pardus*	*Panthera leo*	*Panthera spelaea*	*Panthera leo* sp.
58	Prado Vargas	Cornejo, Burgos						×			
59	Valdegoba	Huérmeces, Burgos	×	×	×			×			×
60	Cueva de la Ermita	Hortigüela, Burgos						×			
61	Caballón	Oña, Burgos	×	×			×	×			
62	La Blanca	Oña, Burgos	×	×							
63	Cueva Millán	Hortigüela, Burgos	×		×						
64	Cueva de Aguilón	Aguilón, Zaragoza						×			
65	Zarzamora-Buho	Perogordo, Segovia						×			
66	Camino	Pinilla del Valle, Madrid						×			
67	Cueva de los Torrejones	Tamajón, Guadalajara						×			

Tab. 2 Archaeological sites at the centre of the Iberian Peninsula and southern side of the Pyrenees, location of the sites and felid species presence.

no.	site	location	*Felis silvestris*	*Lynx pardina*	*Lynx spelaea*	*Lynx lynx*	*Lynx* sp.	*Panthera pardus*	*Panthera leo*	*Panthera spelaea*	*Panthera leo* sp.
68	Cueva de los Casares	Riba de Saelices, Guadalajara	×	×	×		×	×			×
69	Moros de Gabasa	Peralta de Calasanz, Huesca	×		×	×		×			×
70	Cova dels Muricecs	Llimiana, Lleida		×	×			×			
71	Estebanvela	Ayllón, Segovia	×	×							
72	Congosto	S. Andrés del Congosto, Guadalajara	×		×		×				
73	Chaves	Casbas de Huesca, Huesca		×							
74	Gato II	Épila, Zaragoza		×							
75	Jarama VI	Valle del Jarama, Guadalajara									×

Tab. 2 (continued)

no.	site	location	*Felis silvestris*	*Lynx pardina*	*Lynx spelaea*	*Lynx lynx*	*Lynx* sp.	*Panthera pardus*	*Panthera leo*	*Panthera spelaea*	*Panthera leo* sp.
76	Cova de S'Espasa	Oix-Sardenes, Girona						×			
77	Cova dels Ermitons	Sales de Llierca, Girona			×			×			
78	Cova de l'Arbreda	Serinyá, Girona	×	×	×	×		×			×
79	Bora Gran	Serinyá, Girona	×				×				
80	Mollet I	Serinyá, Girona			×	×		×			×
81	Mollet III	Serinyá, Girona			×						
82	Reclau Viver	Serinyá, Girona			×						×
83	Olopte	Isobol, Girona			×	×					
84	Cau del Duc	Torroella de Montgrí, Girona		×	×	×		×			
85	Abric Romaní	Capellades, Barcelona	×	×				×			

Tab. 3 Archaeological sites at the Mediterranean coast, location of the sites and felid species presence.

no.	site	location	*Felis silvestris*	*Lynx pardina*	*Lynx spelaea*	*Lynx lynx*	*Lynx* sp.	*Panthera pardus*	*Panthera leo*	*Panthera spelaea*	*Panthera leo* sp.
86	Cova del Gegant	Sitges, Barcelona		×	×			×			
87	Cau de Coçes	Sitges, Barcelona		×							
88	Musclé	Sitges, Barcelona		×	×						
89	Castell-defels	Castell-defels, Barcelona	×					×			
90	Toixone-res	Toixoneres, Barcelona	×		×						
91	Toll	Moiá, Barcelona	×	×	×	×					×
92	Cingle Vermell	Vilanova de Sau, Barcelona					×				
93	Parco	Alos de Llobregat, Barcelona					×				
94	Cova Fo-radada	Calafell, Tarragona						×			
95	Cova de L'Olla	Marmellar, Tarragona		×		×					
96	Balma de la Griega	Calafell, Tarragona					×				
97	Matu-tano	Vilafamés, castellón	×	×			×				
98	Cueva Horadada	Azuebar, Castellón						×			
99	Cueva de les Mera-velles	Alzira, Valencia						×			
100	Bolomor	Tavernes de Valldigna, Valencia						×			×
101	Mallaetes	Barx, Valencia					×	×			
102	Cova Ne-gra	Xátiva, Valencia	×	×				×			×
103	Cova de Racò del Duc I	Vilallonga, Valencia						×			
104	Parpalló	Gandía, Valencia	×				×				
105	Abrigo de la Queb-rada	Chelva, Valencia					×				
106	Cova del Corb	Ondara, Alicante				×		×			

Tab. 3 (continued)

no.	site	location	*Felis silvestris*	*Lynx pardina*	*Lynx spelaea*	*Lynx lynx*	*Lynx* sp.	*Panthera pardus*	*Panthera leo*	*Panthera spelaea*	*Panthera leo* sp.
107	Cova Foradada	Xàbia, Alicante						×			
108	Cova de les Calaveres	Benidoleig, Alicante						×			
109	Cova del Parat	Tollos, Alicante						×			
110	Cova del Salt	Alcoi, Alicante						×			
111	Abric del Pastor	Alcoi, Alicante						×			
112	Cendres	Teulada-Moraira, Alicante	×				×				
113	Beneito	Muro de Alcoy, Alicante	×				×				
114	Sima de las Palomas	Torre Pacheco, Murcia						×			
115	Cueva del Niño	Ayna, Albacete					×				
116	Cueva Negra	Caravaca de la Cruz, Murcia		×							

Tab. 3 (continued)

no.	site	location	*Felis silvestris*	*Lynx pardina*	*Lynx spelaea*	*Lynx lynx*	*Lynx* sp.	*Panthera pardus*	*Panthera leo*	*Panthera spelaea*	*Panthera leo* sp.
117	Cueva Ambrosio	Vélez Blanco, Almería	×	×			×				
118	Cueva de la Carigüela	Darro, Granada					×	×			
119	Cueva Horá	Darro, Granada		×							
120	Campana de Piñar	Piñar, Granada						×			
121	Cueva de Nerja	Nerja, Málaga	×	×			×				
122	Boquete de la Zafarraya	Alcaucín, Malaga		×				×			
123	Devil's Tower	Gibraltar		×				×			
124	Genista Cave	Gibraltar		×				×			
125	Vanguard Cave	Gibraltar						×			

Tab. 4 Archaeological sites at the southern and western sides of the Iberian Peninsula, location of the sites and felid species presence.

no.	site	location	*Felis silvestris*	*Lynx pardina*	*Lynx spelaea*	*Lynx lynx*	*Lynx* sp.	*Panthera pardus*	*Panthera leo*	*Panthera spelaea*	*Panthera leo* sp.
126	Gorham's Cave	Gibraltar	×				×	×			
127	Gruta do Escoural	Montemor-o-Novo	×		×			×			×
128	Gruta da Figueira Brava	Brava, Setubal	×					×			×
129	Pego do Diabo	Loures	×		×			×			
130	Pedreira da Salemas	Loures	×		×			×			×
131	Salemas	Peniche	×		×						
132	Gruta das Fontaìnhas	Cadaval	×					×			
133	Furninha	Peniche	×		×			×			
134	Casa da Moura	Óbidos						×			
135	Gruta da Oliveira	Torres Novas						×			
136	Gruta do Caldeirao	Tomar	×		×			×			
137	Lorga do Dine	Vinhais						×			×
138	Lapa da Rainha	Leira	×		×						
139	Columbeira	Bombarral	×		×						
140	Casa da Moira	Peniche		×							
141	Algar do Joao Ramos	Peniche		×	×						
142	Algar do Cascais	Lisboa			×						
143	Maltravieso	Cáceres, Extremadura		×							

Tab. 4 (continued)

ANTHONY J. STUART

THE EXTINCTION OF LARGE CARNIVORES IN NORTHERN EURASIA

BACKGROUND

In 1876 the eminent biologist Alfred Russell Wallace observed, with remarkable insight: »we live in a zoologically impoverished world from which all the hugest and fiercest and strangest forms have recently disappeared. It is surely a marvellous fact and one that has hardly been sufficiently dwelt upon, this dying out of so many large Mammalia, not in one place only, but over half the land surface of the globe« (Geographical Distribution of Animals, p. 150).
What killed off the mammoths, woolly rhinos, sabre-tooths, giant ground sloths and so many other spectacular giants (»megafauna«), that thrived on all continents (except Antarctica) during the »Ice Age«; some until only a few thousand years ago? »Megafauna« are usually defined as animals with average body weights of 45 kg or more. We tend to think of these extinct giants as almost fantastic »prehistoric monsters«, but as recognized by Wallace these beasts should still be with us if something drastic and extraordinary had not happened. The greatest losses occurred in South America, North America and Australia, but were also severe in northern Eurasia. Only southern Asia and Sub-Saharan Africa suffered very few losses, so that for example rhinos and elephants still occur there at the present day, although now these iconic survivors are seriously threatened by human activities, including hunting and habitat destruction. Was it humans, climate change, a combination of both that were responsible for their extinction – or perhaps something else? The well-established proposed causes are: unsustainable levels of hunting by humans (known as »overkill«); environmental change, or a combination of both. Each of these has its advocates, but there are serious objections to other proposed explanations such as epidemic disease and an asteroid strike.

MEGAFAUNAL EXTINCTIONS IN NORTHERN EURASIA

For hundreds of thousands of years a succession of spectacular large mammals (»megafauna«) roamed northern Eurasia. For example the rich fossil assemblage from West Runton (North Norfolk / GB), about 700 000 years old, includes: a huge mammoth (*Mammuthus trogontherii*); two species of horse; a rhino; wild boar; five species of deer; and a bison. Carnivores include: lion, spotted hyaena, a bear, »European jaguar« (*Panthera gombaszoegensis*) and sabre-tooth (*Homotherium latidens*), in addition to many small vertebrates. Moving 400 000 years forward in time, the large mammals recovered from the main layer at Schöningen (Lkr. Helmstedt / D), about 300 000 years old, include: humans – as shown by stone tools and the famous wooden spears; two species of horse; two species of rhino; wild boar; two kinds of deer; a bison; aurochs; and sabre-tooth (*Homotherium latidens*). Rather older layers at Schöningen yielded remains of additional species: straight-tusked elephant (*Palaeoloxodon antiquus*), lion and cave bear. Long-term changes to the northern Eurasian fauna involved evolution, extinction of some species, and immigration of new ones, but maintaining a broadly similar range of large animals until about 100 000 years ago (Late Quaternary), after which major extinctions – without replacement by other species – began to occur.

In: Nicholas J. Conard · Henning Hassmann · Kurt Felix Hillgruber · Jordi Serangeli · Thomas Terberger (eds), The *Homotherium* Finds from Schöningen 13II-4. Man and Big Cats of the Ice Age. Forschungen zur Urgeschichte aus dem Tagebau von Schöningen 4 (Mainz 2022). DOI: https://doi.org/10.11588/propylaeum.1006.c13524

The Late Quaternary comprises the Last Interglacial (about 130 000-110 000 years ago and rather warmer than today), through the predominantly cold Last Glacial (interrupted by many warmer interstadials), to the temperate Holocene interglacial (postglacial), which began around 11 700 years ago – and in which we still live. The megafauna of the Last Interglacial (Eemian) in northern Eurasia included: straight-tusked (or forest) elephant (*Palaeoloxodon antiquus*); woolly mammoth (*Mammuthus primigenius*); narrow nosed rhino (*Stephanorhinus hemitoechus*); woolly rhinoceros (*Coelodonta antiquitatis*); steppe bison (*Bison priscus*); giant deer (*Megaloceros giganteus*); extinct camel (*Camelus knoblochi*); Neanderthal (*Homo neanderthalensis*); cave lion (*Panthera spelaea*); cave bears (*Ursus spelaeus, Ursus ingressus, Ursus kudarensis*); and spotted hyaena (*Crocuta crocuta*). All of these went extinct during the Last Glacial, with the exception of woolly mammoth which persisted until 10 700 years (early Holocene) in north-central Siberia, and famously on Wrangel Island (in the Arctic Ocean off north-eastern Siberia) to around 4000 years ago. Spotted hyaena, also disappeared entirely from Eurasia, but still survives today in Africa south of the Sahara.

RADIOCARBON DATING

Establishing an accurate time scale and dating individual animal specimens is crucial to unravelling what happened to the megafauna. Radiocarbon dating is a powerful tool for accurately dating organic remains as far back as more than 40 000 years ago (exceptionally to around 50 000 years). Fortunately, this time range covers most megafaunal extinctions in most parts of the world, including northern Eurasia. The isotope Carbon 14 (^{14}C, radiocarbon) is constantly being produced in the upper atmosphere and absorbed in the form of carbon dioxide into plant tissues. Animals incorporate radiocarbon, along with the much more plentiful stable isotope Carbon 12 (^{12}C), by eating plants or by eating other animals. The unstable ^{14}C undergoes radioactive decay at a constant known rate, but this is continually replenished while the animal is alive. The »clock« starts at the time of death as new ^{14}C is no longer taken in, so that by accurately measuring the remaining ^{14}C the age of the remains of a plant or animal can be calculated. However, by around 50 000 years ago the remaining amount of radiocarbon is so small that it is impossible to measure accurately, which sets an age limit on dating a sample. Collagen (a protein), extracted from bones, teeth and antlers, is the primary material for radiocarbon dating of megafaunal remains. Charcoal and other material found with megafaunal remains can only give us indirect dates which can be misleading when the degree of association is uncertain. In order to reliably date megafauna we need samples of material from the animal itself. The increasingly wide use of AMS (Accelerator Mass Spectrometry) allows very small samples of half a gram or less to be dated accurately. Of course, it is essential that the sample submitted for dating should be correctly identified to species. However, there is a very important complicating factor that needs to be taken into account. Levels of radiocarbon in the atmosphere have varied significantly in the past, which means that measured radiocarbon dates need to be calibrated against a known time scale in order to convert them to calendar years. All of the dates quoted here have been calibrated.

The extensive and careful application of radiocarbon dating has made it possible to build up a picture of megafaunal extinctions in northern Eurasia. The pattern of extinctions (**fig. 1**) is conspicuously staggered over many thousands of years, with some species disappearing before the Last Glacial Maximum, some much later in the Late Glacial, and others surviving into the Holocene or to the present day. This pattern contrasts with North America where most losses seem to have occurred within the Late Glacial. In northern Eurasia, each megafaunal species exhibits a unique and complex pattern of distributional shifts, culminating in extinction for some species and survival in others.

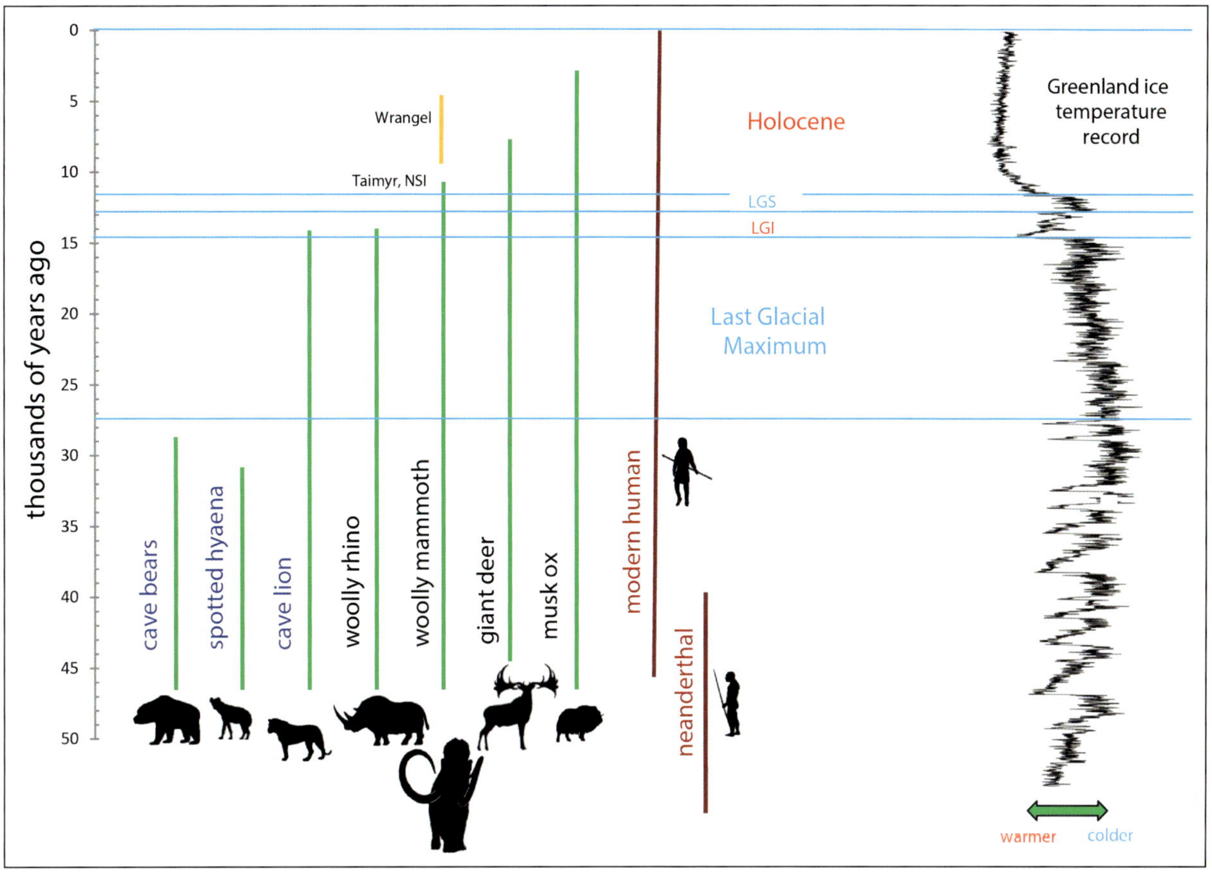

Fig. 1 Chart of extinctions and survivals of representative megafauna during the last 50 000 years in northern Eurasia, based on radiocarbon dates in comparison with the Greenland ice temperature curve. The Last Glacial Maximum was a prolonged cold period; the Holocene a prolonged warm phase in which we still live. Carnivores are lettered in blue. The time ranges of Neanderthals and modern humans are also shown. – LGI: Late Glacial Interstadial (a warm phase); LGS: Late Glacial Stadial (a cold phase); NSI: New Siberian Islands (north-central Siberia); Wrangel: Wrangel Island (north-east Siberia).

LATE QUATERNARY LARGE CARNIVORES

Cave bears (fig. 2)

In the Late Quaternary cave bears (*Ursus spelaeus* and *Ursus ingressus*) were confined to upland areas in western and central Europe, with an isolated population in the Urals. They were absent from northern Europe and also Britain, where only the living brown bear (*Ursus arctos*) occurred. A separate species *Ursus kudarensis* was isolated in the Caucasus. Most tooth wear studies and isotope analyses indicate that cave bears, although belonging to the order *Carnivora*, were largely vegetarian, quite unlike modern brown bears which also eat substantial amounts of animal food. However, isotopic studies indicate that cave bears from Carpathian sites (Romania) enjoyed a broader diet like brown bears today.

The latest of 175 reliable dates (median calibrated years before present) on cave bears (*Ursus spelaeus* and *Ursus ingressus*) are: 28 700 years ago (laboratory number GrA-52632) from Rochedane (dép. Doubs/F); 28 984 years ago (ETH-16879) from Bame aux Pirotas (ct. Jura/CH); and 28 540 years ago (Beta-156100) from Vindija Cave (HR). The latest dates from the Urals indicate that they disappeared from this region around 13 000 years earlier than from central Europe.

Fig. 2 Cave bear skull and mandible (Balkans), total length 50 cm. – (Photo A. J. Stuart).

Spotted hyaena (fig. 3)

In the Last Glacial spotted hyaenas ranged from Iberia and the British Isles, eastwards across central and southern Europe, to European Russia, southern Siberia, and the Pacific coast in the Russian Far East. They were entirely absent north of 58 degrees latitude. Spotted hyaenas in Africa at the present day are active pack (clan) hunters of large ungulates up to the size of wildebeest and zebra, as well as scavengers of their carcasses. The powerful jaws and massive teeth have evolved for specialized bone crushing, thus allowing the maximum nutrition to be extracted from a carcass. Likely prey species in Last Glacial northern Eurasia include adult and juvenile horse, reindeer, red deer, giant deer (*Megaloceros giganteus*), extinct bison (*Bison priscus*), aurochs (*Bos primigenius*) and wild boar. No doubt hyaenas would have disputed »kills« with humans.

On the basis of 99 dates, spotted hyaena is last recorded at 30 813 years ago (OxA-10523) from Paglicci Cave (prov. Foggia/I), and 31 433 years ago (GrA-2812) from Goyet Cave (prov. Namur/B). The latest dates from central Europe to the Russian Far East are more than 9000 years older. The available evidence indicates that spotted hyaena disappeared from northern Eurasia probably in response to the onset of the Last Glacial Maximum, due to cooling climate and also reduction of its herbivore prey abundance driven by deteriorating vegetational productivity. Spotted hyaenas probably disappeared from central Europe and eastwards about 40 000 years ago, but persisted substantially later – to about 31 000 years ago – in north-western and southern Europe where the climate remained milder for longer.

Fig. 3 Partial left mandible of spotted hyaena (Urals, Russia). – (Photo A. J. Stuart). – Scale 1:2.

Fig. 4 Cave lion right mandible (Lena River, north-eastern Siberia). – (Photo A. J. Stuart). – Scale 1:2.

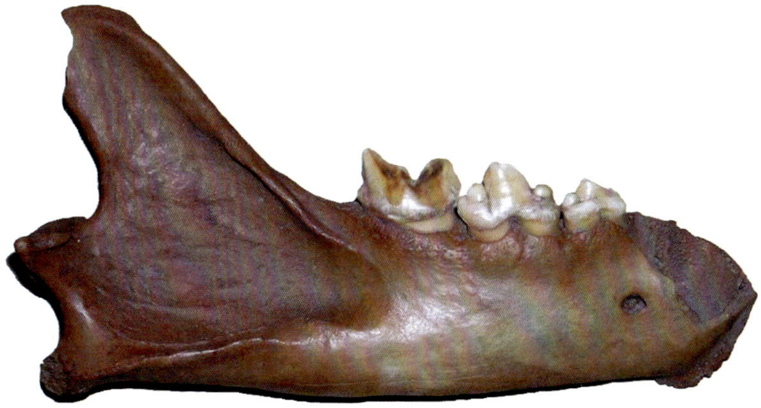

Cave lion (fig. 4)

During the Late Quaternary the cave lion, *Panthera spelaea* (an extinct species distinct from modern lions) was widespread across northern Eurasia to north-eastern Siberia and was able to cross the Bering region into north-western North America (Alaska and the Yukon). Cave lions were rather larger than modern lions and probably could have tackled larger prey. In northern Eurasia the likely prey species would have included horse, reindeer, giant deer, red deer, musk ox, extinct bison, and occasionally young woolly rhino and young mammoth.

The latest of 140 dates that we have for cave lion are: 14378 years ago (OxA-17268) from Zigeunerfels Cave (Lkr. Sigmaringen/D; **fig. 5**); 14141 years ago (AA-41882) from Le Closeau (dép. Seine-et-Marne/F); and 14640 years ago (OxA-12901) from the Lena River in north-eastern Siberia. Younger dates from Alaska suggest that they persisted about 900 years later in that region. This evidence indicates that cave lion extinction occurred within the Late Glacial period throughout Eurasia by about 14000 years ago, very likely in response to a warming climate accompanied by the spread of trees and shrubs. Perhaps there was also a reduction in numbers of available prey, although nearly all its potential prey species survived much later. There is no indication that they survived anywhere into the Holocene (i.e. later than 11700 years ago).

Fig. 5 Cave lion upper canine tooth (Zigeunerfels Cave, southern Germany). – (Photo W. von Koenigswald). – Scale 1:2.

Sabre-tooth

The single radiocarbon-dated find of sabre-tooth (*Homotherium latidens*), based on a mandible trawled from the North Sea, is discussed elsewhere in this publication. My view is that we need more evidence to demonstrate its presence in Europe during the Last Glacial. Otherwise the latest record is much older – from Schöningen at around 300000 years ago.

LARGE CARNIVORES AND PALAEOLITHIC HUMANS

The presence of formidable large carnivores throughout the Quaternary no doubt posed a constant danger to humans as well as disputing their kills, although in turn humans very likely stole carnivore kills when the opportunity arose. As has been suggested, the 300 000-year-old Schöningen spears are likely to have been used in defence against predators, including sabre-tooths, lions and hyaenas, as well as for hunting large game.

Modern humans arrived in Europe about 45 000 years ago, via the Middle East, replacing the Neanderthals, who preceded them within a few thousand years. However, by far, the majority of extinctions in northern Eurasia, including the large carnivores, occurred much later – which is inconsistent with »overkill«. The evidence that we have so far favoured the idea that carnivore extinctions were driven by climatic and/or vegetational changes. However, the possible role of humans, especially in reducing remaining populations to the point of extinction, has yet to be satisfactorily explored.

There are many artistic depictions of cave lions, including the beautiful multiple painted images from Chauvet Cave (dép. Ardèche/F) and the enigmatic lion-headed human figure (carved out of mammoth tusk) from Hohlenstein Stadel (Alb-Donau-Kreis/D). Although lions were clearly prominent in the minds of Upper Palaeolithic people, there is no evidence that they were hunted. So far, depictions of cave bears, with their distinctive domed foreheads, are known only from Chauvet Cave. However, numerous cut marks on cave bear bones from the cave sites Casamène (dép. Doubs/F) and Hohle Fels (Alb-Donau-Kreis/D) clearly show that their carcasses were exploited by humans, although direct evidence suggesting that they were hunted by people is extremely scarce. The only example known so far is from Hohle Fels Cave, where a bear thoracic vertebra (either cave bear or brown bear) was found with a small triangular fragment of *Hornstein* embedded in it; perhaps broken off a much larger projectile point. In the case of hyaenas, there are no convincing representations from any site and so far no evidence has been found to suggest that any were killed by humans.

CONCLUDING REMARKS

So radiocarbon dates (as well as stratigraphic evidence) from many localities show a large difference between the histories of cave bears and spotted hyaena on the one hand and cave lion on the other. Cave lion survived into the Late Glacial period about 14 000 years ago, whereas cave bears and spotted hyaena disappeared around 13 000 years earlier, before the beginning of the Last Glacial Maximum about 27 500 years ago. Significantly, there is no evidence whatsoever that the extinction of large carnivores resulted from the extinction of their herbivore prey. It should be pointed out that for each species the actual time of extinction is likely to be a little later than the youngest available date. Further advances in our understanding of megafaunal extinctions will involve obtaining many more radiocarbon dates as well as the increasingly important developments in isotope chemistry and ancient DNA.

ACKNOWLEDGEMENTS

I am most grateful for the opportunity to attend the excellent and stimulating workshop at the Paläon Museum in June 2015. Especial thanks are due to my principal collaborator Adrian Lister and the UK Natural Environment Research Council who have funded our research.

REFERENCES

Barnosky et al. 2004: A. D. Barnosky / P. L. Koch / R. S. Feranec / S. L. Wing / A. B. Shabel, Assessing the causes of Late Pleistocene extinctions on the continents. Science 306, 2004, 70-75.

van Kolfschoten/Buhrs/Verheijen 2015: T. van Kolfschoten / E. Buhrs / I. Verheijen, The larger mammal fauna from the Lower Paleolithic Schöningen spear site and its contribution to hominin subsistence. Journal of Human Evolution 89, 2015, 138-153.

von Koenigswald 2002: W. von Koenigswald, Lebendige Eiszeit; Klima und Tierwelt im Wandel (Darmstadt 2002).

Pacher/Stuart 2009: M. Pacher / A. J. Stuart, Extinction chronology and palaeobiology of the cave bear *Ursus spelaeus*. Boreas 38, 2009, 189-206.

Reumer et al. 2003: J. W. F. Reumer / L. Rook / K. van der Borg / K. Post / D. Mol / J. de Vos, Late Pleistocene Survival of the Saber-toothed Cat *Homotherium* in Northwestern Europe. Journal of Vertebrate Paleontology 23/1, 2003, 260-262.

Stiller 2014: M. Stiller, Mitochondrial DNA diversity and evolution of the Pleistocene cave bear complex. Quaternary International 339-340, 2014, 224-231.

Stuart 2015: A. J. Stuart, Late Quaternary megafaunal extinctions on the continents: a short review. Geological Journal 50, 2015, 338-363.

Stuart/Lister 2010: A. J. Stuart / A. M. Lister, The West Runton Freshwater Bed and the West Runton Mammoth: Summary and Conclusions. Quaternary International 228/1-2, 2010, 241-248.

2011: A. J. Stuart / A. M. Lister, Extinction chronology of the cave lion *Panthera spelaea*. Quaternary Science Reviews 30, 2011, 2329-2340.

2012: A. J. Stuart / A. M. Lister, Extinction chronology of the woolly rhinoceros *Coelodonta antiquitatis* in the context of late Quaternary megafaunal extinctions in northern Eurasia. Quaternary Science Reviews 51, 2012, 1-17.

2014: A. J. Stuart / A. M. Lister, New radiocarbon evidence on the extirpation of the spotted hyaena (*Crocuta crocuta* (Erxl.)) in northern Eurasia. Quaternary Science Reviews 96, 2014, 108-116.

2021: A. J. Stuart, Vanished Giants; the Lost World of the Ice Age (Chicago 2021).

SUMMARY / ZUSAMMENFASSUNG

The Extinction of Large Carnivores in Northern Eurasia

During the Late Quaternary northern Eurasia supported a range of large mammals (megafauna) that are now extinct or extirpated from the region. These include cave bears (*Ursus spelaeus* and *Ursus ingressus*), spotted hyaena (*Crocuta crocuta*) (still lives in Africa), cave lion (*Panthera spelaea*), and possibly also a sabre-tooth cat (*Homotherium latidens*). Radiocarbon dates made directly on megafaunal remains allow us to estimate when each disappeared. The dates quoted here have all been calibrated, that is converted into calendar years. Cave bears and hyaenas had all disappeared before the Last Glacial Maximum 27 500 years ago, whereas cave lion survived much longer until about 14 000 years ago.

Das Aussterben der großen Fleischfresser in Nordeurasien

Während des späten Quartärs gab es in Nordeurasien eine Reihe von großen Säugetieren (Megafauna), die heute ausgestorben oder in der Region ausgerottet sind. Dazu gehören Höhlenbären (*Ursus spelaeus* und *Ursus ingressus*), Tüpfelhyänen (*Crocuta crocuta*) (diese leben heute noch in Afrika), Höhlenlöwen (*Panthera spelaea*) und möglicherweise auch eine Säbelzahnkatzenart (*Homotherium latidens*). Anhand von ^{14}C-Daten, die direkt an den Überresten der Megafauna vorgenommen wurden, lässt sich abschätzen, wann diese verschwunden sind. Die hier angegebenen Daten wurden alle kalibriert, d. h. in Kalenderjahre umgerechnet. Höhlenbären und Hyänen waren bereits vor dem letzten glazialen Maximum vor 27 500 Jahren verschwunden, während der Höhlenlöwe noch bis vor etwa 14 000 Jahren überlebte.

ART

KURT FELIX HILLGRUBER · THOMAS TERBERGER

BIG CATS AND HUMANS – SOME REMARKS ON THE INTERACTION WITH BIG CATS IN PREHISTORIC ART

Today, over 400 million cats accompany us in the western world and they are our favorite pets, rivaled only by dogs. Our close companionship with cats is the result of a long development. When felids were first domesticated is unknown, but cat bones on the island of Cyprus provide important information: at the settlement site Klimonas cat remains were identified in building 1, which is dated to the Pre-Pottery Neolithic (PPN A; *c.* 8800-8600 cal BC). Due to the lack of a wild form of cat present on the island, the recovered remains were undoubtedly brought to the island over the sea (Vigne et al. 2012). Further evidence for a more prominent role of cats during the PPN is available from sites such as Jericho (Zeuner 1958).

Similar to the domestication of the wolf, the first cat remains found in archaeological assemblages were from wild animals presumably killed for their furs and found subsequently together with hunted remains of other carnivores like fox, lynx, bear or wolf. In contrast to the pack of wolfs, who are believed to have followed the campsites of men, the theory behind the domestication of cats is probably tied together with the hunting of adult cats and the subsequent discovery of orphaned kittens, who stayed in the care of humans. The early remains of cats found in the Neolithic show that these animals may have already been valued human companions, but it was still a wildcat belonging to the *lybica* race.

However, mankind's fascination for cats is much older than its domestication of a companion and their use as an aid against vermin in Neolithic societies. Our hunter-gatherer ancestors already held the large Pleistocene cats in high esteem. Regularly occurring depictions in the Ice Age cave art are known throughout Western Europe with over 120 drawings (Serangeli 2006, 45). In part this fascination with the big cats is rooted in the fact that for a long period of our evolution we were the hunted and not the hunter. Early evidence for a big cat victim is the calvarium from Swartkrans, South Africa, with two perforations (**fig. 1**).

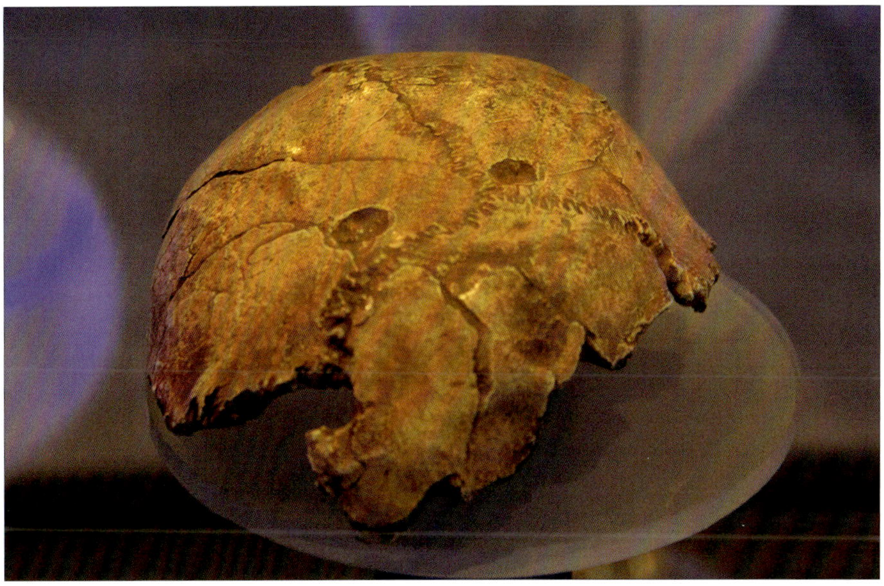

Fig. 1 Hominin skull from Swartkrans with the two perforations, probably caused by *Dinofelis*. – (Photo M. Peel, Creative Commons).

In: Nicholas J. Conard · Henning Hassmann · Kurt Felix Hillgruber · Jordi Serangeli · Thomas Terberger (eds), The *Homotherium* Finds from Schöningen 13II-4. Man and Big Cats of the Ice Age. Forschungen zur Urgeschichte aus dem Tagebau von Schöningen 4 (Mainz 2022). DOI: https://doi.org/10.11588/propylaeum.1006.c13525

The puncture marks in the parietal bone and their distance matched the spacing of the canines of a fossil sabre-toothed cat of the genus *Dinofelis,* the skull of which was found in the same part of the cave. It is reasonable to assume that the young *Australopithecus africanus* was killed by the sabre-toothed cat and that it was carried off to the feeding place in the cave (Brain 1969; 1981). Another example for a hominin killed by a large carnivore was discovered at the famous site of Dmanisi in Georgia: At the base of skull 2282 two perforations were found, which may indicate that the small *Homo georgicus* perhaps fell prey to the *Megantereon,* also present in the fossil assemblage of the site (Johanson/Wong 2010).

VICTIMS BECOME HUNTERS?

After the examples of our ancestors being prey the question arises: when did humans reach a stage of cultural development in which they were not only prey but could compete with large predators? Here, finds from the last decades provide interesting insights. The discovery of the Schöningen spears in 1994 and 1995, the oldest complete, unambiguous hunting weapons of mankind, as a clear material sign of the ability of our ancestors to compensate their physical deficits with innovations. The question arises whether these weapons were not also used for defensive purpose (Serangeli/van Kolfschoten/Conard 2014).

Basically, finding carnivore remains in Middle Pleistocene sites is nothing unusual. Many carnivores regularly used caves as a refuge or to raise their young (Sutcliffe 1970; Fosse 1996; Graham et al. 2013). Thus, for archaeological sites, especially in caves, the presence of carnivore remains must be examined in order to differentiate between three different find circumstances: First of all, the carnivores as accumulators and the use of the site as a den; secondly, the carnivores as scavengers of for example hominid waste; and lastly, and more difficult to assess, carnivores as hominids' prey. Already in the Middle Pleistocene, an indirect reference to the changing relationship between hominins and large predators can again be concluded through a find from Schöningen (Lkr. Helmstedt/D). In the immediate vicinity of the spears, a humerus of a sabre-toothed cat was found with scrape and percussion marks in 2013 (see contribution Verheijen et al. in this volume). Similar single finds of carnivore bones with traces of human manipulation are known throughout Europe, ranging from cutmarks on bear bones from Boxgrove (West Sussex/GB; Roberts/Parfitt 1999, 398) or Taubach (city of Weimar/D; Bratlund 1999, 113) to cutmarks on cave lion bones from the Gran Dolina at Atapuerca (prov. Burgos/E; Blasco et al. 2010) or a burned and worked fragment of a leopard bone from Cova Negra (prov. Valencia/E; Sanchis et al. 2015). The late Middle Palaeolithic open air site of Salzgitter-Lebenstedt (Lower Saxony) provided two well-preserved lion remains: a metacarpus and a canine. No clear cut marks are visible on the finds but they are probably the results of Neanderthal selection and the canine might have been used as a pendant (Staesche 2017).

The finds mentioned prove the long-lasting evidence for the use of predator bodies to obtain resources. The cut marks on the bones might be related to hunted individuals. By the beginning of the Upper Palaeolithic, the relationship with big cats becomes more important and is expressed in different ways. First of all, the use of hides and the acquiring of meat inferred through cut marks on bones of carnivores, secondly the use of carnivores bones as tools and finally ornament production from carnivore teeth and depicting them in art. The first example is well documented in carnivore exploitation in Western Europe like the Swabian Jura (Camarós et al. 2015) or Central European sites like Pavlov (Moravia/CZ; Wojtal et al. 2020), even though predominantly cave bears, wolves and foxes, and very rarely lions and hyenas were exploited (Camarós et al. 2015, 4). The same holds true for bone tools. Some beautiful retouchers made from lion canines were discovered in the Hohle Fels (Alb-Donau-Kreis/D) and Vogelherd (Swabian Jura/D) (Kitagawa et al. 2012).

Fig. 2 Ivory felid figurine from the Vogelherd Cave on the Swabian Alb. – (Photo J. Lipták, University of Tübingen).

While the Upper Palaeolithic is characterized by the increasing abundance of ornaments and art, felid remains used as raw material are uncommon. Although the use of lion remains by Upper Palaeolithic people lacks those of other carnivores, they occupy a special position in Palaeolithic art.

BIG CATS AND ART – SOME LESSONS TO LEARN

»Power and aggression« (Kraft and Agression) was the title of a publication by late Joachim Hahn, specialist on early art (Hahn 1986). He noticed the prominent role of wild animals in general and big cats in special among the sculptures made of mammoth ivory found in the Aurignacian levels of the well-known caves of Vogelherd and Hohlenstein-Stadel in the Lone Valley and the Geissenklösterle and Hohle Fels in the Ach Valley on the Swabian Alb (Alb-Donau-Kreis/D). The *c.* 50 figurines belong to the most ancient finds of Palaeolithic art, dating to about 35 000-32 000 cal BC (Conard/Bolus 2003, 336). The first sculptures discovered in 1931 in the Vogelherd Cave by Gustav Riek, indeed depicted carnivores like lions and bears (Riek 1932) and persuaded Hahn to postulate his message and stressing the importance of these powerful animals (**fig. 2**).
In the last decade, renewed excavations at the Vogelherd Cave led to the discovery of a sizable number of figurines depicting smaller animals like fish, birds or herbivores. They contribute to a more varied and balanced image of animals (Floss 2018). In conclusion, the sources of inspiration for the artists were not only dangerous species, but a broader spectrum of animals with very different attributes. The elegant animals carved in perfect craftmanship may indicate strength, but no aggression is clearly visible. This is further emphasized if we take into account the interpretations of the hybrids – in part humans, in part animals – like the Lion-man of Hohlenstein-Stadel or the adorant from the Geissenklösterle (Wehrberger 2013). The relatively large Lion-man was not a pendant to be carried around and the hypothesis that it was placed at the back of the cave, removed from the light of day, as a »stationary« object of art seems plausible. These pieces may indicate a metaphysical state, which is often referred to as belonging to shamanism (Zeeb-Lanz/Rey-

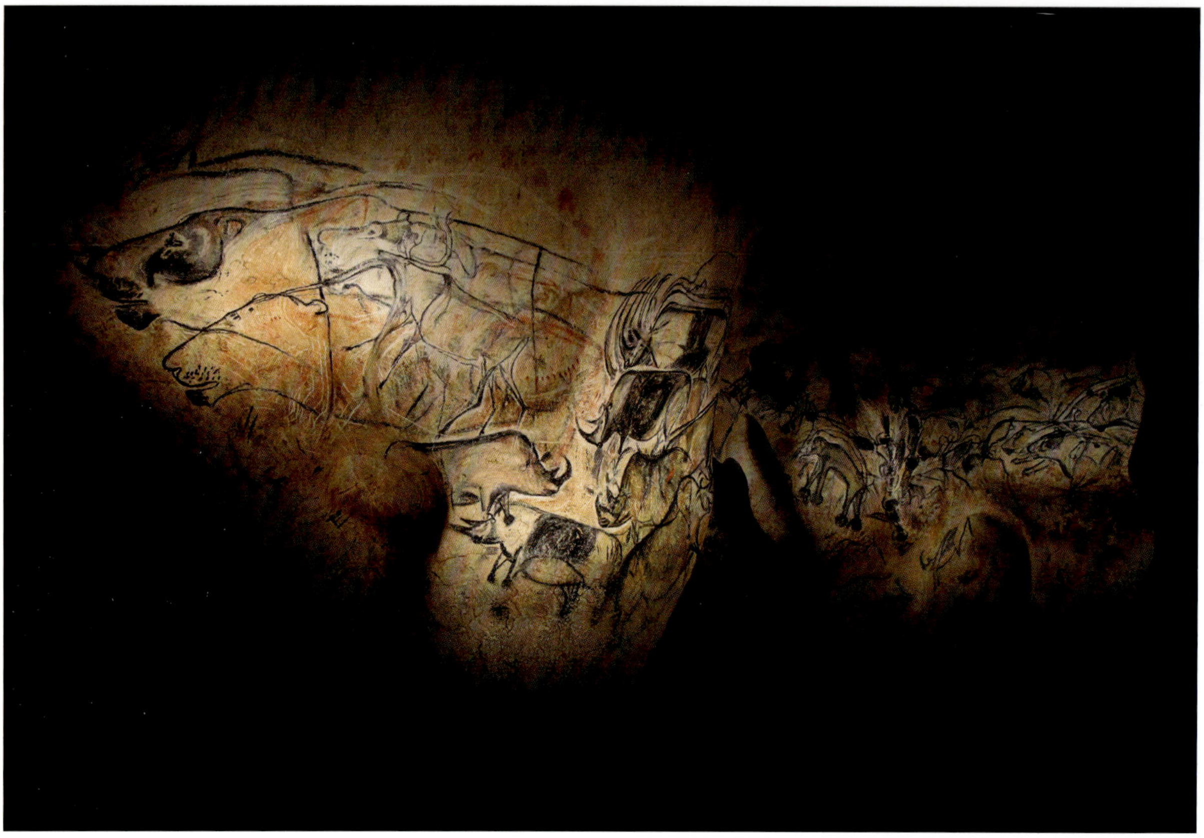

Fig. 3 Lion panel in the Grotte Chauvet. – (Photo P. Aventurier, Caverne-du-pont-darc).

mann 2019). The transcendence of man to an animal form is not a state of mind that is generally associated with aggression or violence. However, there is little doubt on the subject of power in early art, which is also reflected in the animal tooth pendants, due to the choice to often use the canines of carnivores.

The prominent role of big cats in the earliest phase of art is corroborated by the Grotte Chauvet (dép. Ardèche/F) in the Ardèche Valley (**fig. 3**). Discovered in 1994, the Grotte Chauvet yields a fascinating body of early cave art and an outstanding number of drawings of large felids (Clotte 2003). The engravings and paintings on the walls of the cave are unrivalled. Direct dating of the Grotte Chauvet raised considerable scientific discussion (Züchner 1996; Pettitt/Bahn 2003; 2015). However, a more recent evaluation of the radiocarbon evidence suggests two phases of occupation (37 000-33 500 years ago and 31 000-28 000 years ago) and the paintings are definitely older than 28 000 years (Quilesa et al. 2016). In conclusion we perceive a mastery of composition and perspective from the very beginning of cave art. At the same time this demonstrates that there is no evolution of art and no perceived trend towards more elaborate and skillfully applied paintings during in the Pleistocene.

There are further examples of felid representations in cave art and a famous example is from the site of Lascaux in the Dordogne. The images of Lascaux are partially painted, drawn and there are engraved figures done with a high degree of skill. Nearly 600 animals were discovered. Again the herbivores like bison, reindeer, mammoths or ibex dominate the drawings of horses alone make up nearly one quarter. Carnivores, bears and lions, are few in number. In the case of Lascaux one can argue that the carnivores were of special interest to the painters: all of them are placed in the most remote parts of the widely branched cave. Even though, there is no clear trend in the cave art of Western Europe of carnivores purposely being situated in the darker, more removed areas of the caves (Leroi-Gourhan 1973, 569). Some incredible examples of cave

Fig. 4 Felids (bottom center) from the Chambre de Felines in Lascaux who are probably wounded by projectile. – (After Leroi-Gourhan et al. 1979, 327).

art can only be seen after an arduous trip through narrow spaces. Interestingly, one of the most important ivory figurines, the Lion-man from the Swabian Alb was also discovered at the far end of the cave. During the re-examination from 2009 to 2013 the original sediments were found and due to the recovery of some *in situ* ivory splinters, the position of the Lion-man in 1939 could be affirmed (Beutelspacher/Kind 2013).
What kind of felids are depicted in Pleistocene cave paintings is often unclear. Some attributes, like the hair tuft at the end of the tail and the drawings of a group of felids, for example in the Grotte Chauvet or La Vache, may indicate lions which are the sole modern felid species that lives in groups i.e. a pride of lions (Hunter 2015, 11).
The depiction of large felids in the cave art of Western Europe within the Upper Palaeolithic is not as numerous as prey animals like bovids and horses but occurs regularly. The same holds true in later periods. For example, Mary Leakey showed in the Tanzanian rock art in Kandanoa that of 474 depicted animals, 12 % were carnivores (Leakey 1983), mostly lions and hyaenas. Similar observations can be made for historic Massai and Bushman rock art, where lions are depicted even in confrontation with the human hunters, but are way behind the large number of depicted cattle and herbivores (Kruuk 2002, 196; Vinnicombe 1976).
In contrast to the representation of carnivores in classical and historical art, for example in Roman mosaics, the Pleistocene depictions of felids are often not aggressive. The animals are mostly serene, walking or in the example of the final chamber of the Grotte Chauvet where the depiction of a pride is in the process of stalking the prey or drinking in a line at a waterhole. The aggressive act of hunting, especially the moment of killing the prey, is not depicted.
Due to the rare depiction of humanoid figures in Pleistocene cave art one could expect that a direct confrontation between man and felid is absent, but again Grotte Lascaux has a very telling example. Lines or bundles of lines are in direct contact with animals (**fig. 4**). These lines may be interpreted as arrows or spears wounding the prey, especially in conjunction with wiggly lines in front of the animals head. The later may be interpreted as blood spurting from the mortally wounded prey's mouth. A depiction of lions in the cham-

Fig. 5 Engravings of a lions from the Grotte des Trois-Frères. – (After Bégouën et al. 2014, 109).

bre of felins shows a number of arrows pointing at their mid section, while one lion is »bleeding« from its muzzle quite heavily (Leroi-Gourhan et al. 1979).

Even more obvious in its evidence is an engraving of a lion from the Grotte des Trois-Frères (dép. Ariège/F; Bégouën et al. 2014), possibly more than one individual due to a well-made perspective, indicating a number of lions behind each other, with arrows clearly striking its torso (**fig. 5**).

In addition to the fascination we feel when perceiving Pleistocene art, it provides us with the unique chance to see extinct animals through the eye of the beholder. As mentioned above, the depicted art may give us insights into the appearance of life forms surpassing our knowledge derived from fossil finds. For example today's lion, in contrast to other big cats, has a hair tuft at the tip of its tail. Clearly visible on an engraving of a possible lion from the site of Laugerie-Basse (dép. Dordogne/F; **fig. 6**). The lack of a mane in some of the paintings may indicate only female lions or the complete lack of a mane for Pleistocene male cave lions. The development of a mane in modern-day lions is a diverse topic. Climate, regional variation and behavior may be responsible (Yamaguchi et al. 2004). The African lion generally has a more developed mane than its Asian counterpart, with the exception of prides living in open and dry habitats. Male lions in the hot and extremely dry Tsavo National Park usually do not sport a mane. This observation is backed by the heavier manes of lions from European and North American zoos with their cooler ambient temperatures (West/Packer 2003).

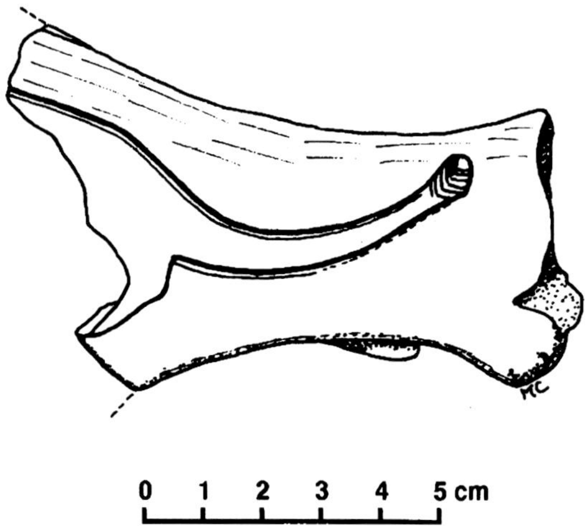

Fig. 6 Engraving of a possible lion with a hair tuft at the tail from the site of Laugerie-Basse. – (After Crémades/Laville 1995, 259).

Fig. 7 Ivory felid figurine from the Vogelherd Cave on the Swabian Alb with x-markings. – (Photo J. Lipták, University of Tübingen).

Especially the later observation may indicate that a mane would be expected in Pleistocene cave lions, but up to now no mane is visible on the drawn figures in Western European rock art. The argument that only female lions are depicted is possible, because the brunt of hunting carried out by prides is borne by the female members, even though male lions are needed in bringing down very large prey like Cape buffalos, giraffes and even elephants. If we take the small ivory statuettes of the Swabian Alb into account, some additional details can be seen. Again no full mane is visibly carved, even though on the small lion head from the Vogelherd a line of x-markings is situated right at the position where modern-day tigers and lynxes sport their typical mutton chops. Similar x-markings are depicted on the famous horse from the Vogelherd, where one would expect its mane. Are the x-markings a possible mane? Two additional figurines of felids from the Vogelherd and the Hohle Fels Cave carry these x-markings. They are not restricted to the shoulder area, where one would expect the mane, but are present along the whole dorsal line. It would appear that these marks do not represent a mane, but perhaps discolorations of the fur like the eel back, the darkly colored stripe along the back (**fig. 7**). Again the Lion-man from the Swabian Alb and the realistic drawings from the Grotte des Trois-Frères can be consulted to ponder the question regarding cave lions and the absence of a mane. First of all, the drawing of a felid at Les Trois-Frères shows a detailed view of the frontal features: Around the face the vibrissae are drawn in a naturalistic way, but no mane is visible. Does this painting show a lioness? The question of gender can be solved by looking at the figurine of the Lion-man from Hohlenstein-Stadel Cave. For a long time the exact gender of the figurine was unclear. By the detection of new fragments and the reassembling of the figurine the crotch can now be identified. A polished ivory plate is visible and this part is offset from the lion. Because of this detail interpreted as the sex, the lion is interpreted as a male individual (Ebinger-Rist et al. 2013, 11).

If the Pleistocene artists were such astute observers of their world, then it is not surprising that not only anatomical attributes but animal behavior is reflected in their drawings. The occurrence of prides of Pleistocene lions can be observed on the perspective drawings of 16 heads in the Grotte Chauvet (Clottes 2003).

Fig. 8 Comparison of Pleistocene art and modern-day behavior of felids. – (a after Clottes 2003, 131; b photo Alamy Stockfoto).

An additional interesting motive is found on the left wall of the end chamber. Two silhouettes of felids are drawn next to each other (**fig. 8**). There is a distinct difference in size, the larger one with a marked scrotum, and both seem to duck and rub each other. This drawing reminds one of the actual mating process of lions, where the larger male is herding the lioness. Another exciting image in the Grotte Chauvet from the Alcove depicts the head of two felids, where the upper one is snarling and the other one is taking on a submissive position with a cowered head, avoiding any eye contact and sporting flattened ears. These behavioral and reproduced motions are not only represented in drawn art but are for example also present on a small ivory plaque from the site of Pavlov, where a felid is stretching in a lazy pose familiar to all who have their own house cat (Woijtal et al. 2020, 11).

Pleistocene mobile and cave art of felids provides important insights into the interaction of humans and felids, as well as some interesting details of the Late Pleistocene big cats and their behaviour. The depictions are sometimes not unambiguous but invaluable in order to reconstruct the appearance of these extinct animals. Art allows a fascinating insight into the thoughts of our ancestors, even though our understanding remains fragmented.

REFERENCES

Bégouën et al. 2014: R. Bégouën / J. Clottes / V. Feruglio / A. Pastoors, La caverne des Trois-Frères: anthologie d'un exceptionnel sanctuaire préhistorique (Paris 2014).

Beutelspacher/Kind 2012: T. Beutelspacher / C.-J. Kind, Zum Fortgang der Untersuchungen in der Stadelhöhle im Hohlenstein. Archäologische Ausgrabungen in Baden-Württemberg 2011 (2012), 66-72.

Blasco et al. 2010: R. Blasco / J. Rosell / J. L. Arsuaga / J. M. Bermúdez de Castro / E. Carbonell, The hunted hunter: the capture of a lion (*Panthera leo fossilis*) at the Gran Dolina site, Sierra de Atapuerca, Spain. Journal of Archaeological Science 37, 2010, 2051-2060.

Bratlund 1999: B. Bratlund, Taubach revisited. Jahrbuch des RGZM 46, 1999, 61-174.

Camarós et al. 2016: E. Camarós / S. C. Münzel / M. Cueto / F. Rivals / N. J. Conard, The evolution of Paleolithic hominin-carnivore interaction written in teeth: Stories from the Swabian Jura (Germany). Journal of Archaeological Science: Reports 6, 2019, 798-809.

Clottes 2003: J. Clottes, Chauvet Cave – The Art of Earliest Times (Salt Lake City 2003).

Conard/Bolus 2003: N. J. Conard / M. Bolus, Radiocarbon dating the appearance of modern humans and timing of cultural innovations in Europe: new results and new challenges. Journal of Human Evolution 44, 2003, 331-371.

Crémades/Laville 1995: M. Crémades / H. Laville, Le félin gravé de Laugerie-Basse: à propos du mouvement dans l'art paléolithique. Paléo 7/1, 1995, 259-265.

Ebinger-Rist et al. 2013: N. Ebinger-Rist / C.-J. Kind / K. Wehrberger / S. Wolf, Von Kopf bis Fuss – Der Löwenmensch näher betrachtet. In: Wehrberger 2013, 64-74.

Hahn 1986: J. Hahn, Kraft und Aggression: die Botschaft der Eiszeitkunst im Aurignacien Süddeutschlands? Archaeologica Venatoria 7 (Tübingen 1986).

Hunter 2015: L. Hunter, Wild Cats of the World (London 2015).

Floss 2018: H. Floss, Same as it ever was? The Aurignacian of the Swabian Jura and the origins of Palaeolithic art. Quaternary International 491, 2018, 21-29.

Fosse 1996: P. Fosse, La Grotte n° 1 de Lunel-Viel (Hérault, France): repaire d'hyenes du Pleistocene Moyen. Etude taphonomique du matériel osseux. Paléo 8, 1996, 47-81.

Graham et al. 2013: R. W. Graham / E. L. Lundelius / L. Meissner / K. Muhlestein, Friesenhahn Cave: Late Pleistocene paleoecology and predator-prey relationships of mammoths with an extinct scimitar cat. Field Guides 30, 2013, 15-31.

Johanson/Wong 2010: D. Johanson / K. Wong, Lucy's Legacy – The Quest for Human Origins (New York 2010).

Kitagawa et al. 2012: K. Kitagawa / P. Krönneck / N. J. Conard / S. C. Münzel, Exploring cave use and exploitation among cave bears, carnivores and hominins in the Swabian Jura, Germany. Journal of Taphonomy 10/3-4, 2012, 439-461.

Kruuk 2002: H. Kruuk, Hunter and Hunted. The Relationship between Carnivores and People (Cambridge 2002).

Leakey 1983: M. Leakey, Africa's Vanishing Art – The Rock Paintings of Tanzania (New York 1983).

Leroi-Gourhan 1973: A. Leroi-Gourhan, L'Art de la Préhistoire (Paris 1973).

Leroi-Gourhan et al. 1979: A. Leroi-Gourhan / J. Allain / G. Delluc / M. Girard / J. Evin / A. Laming-Emperaire / A. Bocquet / R. Bouchez / A. Cornu / J. M. D. Coey / F. H. Schweingruber, Lascaux inconnu. Gallia Prehistoire: Supplement 12 (Paris 1979).

Pettitt/Bahn 2003: P. Pettitt / P. Bahn, Current problems in dating Palaeolithic cave art: Candamo and Chauvet. Antiquity 77, 2003, 134-141.

2015: P. Pettitt / P. Bahn, An alternative chronology for the art of Chauvet cave. Antiquity 89/345, 2015, 542-553.

Quilesa et al. 2016: A. Quilesa / H. Valladas / H. Bocherens / E. Delqué-Koli / E. Kaltnecker / J. van der Plicht / J.-J. Delannoyh / V. Feruglioi / C. Fritz / J. Monneyh / M. Philippe / G.Tosello / J. Clottes / J.-M. Geneste, A high-precision chronological model for the decorated Upper Paleolithic cave of Chauvet-Pont d'Arc, Ardèche, France. Proceedings of the National Academy of Sciences of the United States of America 113, 2016, 4670-4675.

Riek 1932: G. Riek, Paläolithische Station mit Tierplastiken und menschlichen Skelettresten bei Stetten ob Lontal. Germania 16, 1932, 1-8.

Roberts/Parfitt 1999: M. B. Roberts / S. A. Parfitt, Boxgrove A Middle Pleistocene hominid site at Eartham Quarry, Boxgrove, West Sussex. Archaeological Reports N. S. 17 (London 1999).

Sanchis et al. 2015: A. Sanchis / C. Tormo / V. Sauqué / V. Sanchis / R. Díaz / A. Ribera / V. Villaverde, Pleistocene leopards in the Iberian Peninsula: New evidence from palaeontological and archaeological contexts in the Mediterranean region. Quaternary Science Reviews 124, 2015, 175-208.

Serangeli 2006: J. Serangeli, Verbreitung der großen Jagdfauna in Mittel- und Westeuropa im oberen Jungpleistozän. Ein kritischer Beitrag. Tübinger Arbeiten zur Urgeschichte 3 (Rahden/Westf. 2006).

Serangeli/van Kolfschoten/Conard 2014: J. Serangeli / T. van Kolfschoten / N. J. Conard, 300.000 Jahre alte Funde einer Säbelzahnkatze aus Schöningen – Die gefährlichste Raubkatze der Eiszeit erstmals für Norddeutschland belegt. Berichte zur Denkmalpflege in Niedersachsen 2014/1, 10-12.

Staesche 2017: U. Staesche, Der Höhlenlöwe (*Panthera leo spelaea*) aus den mittelpaläolithischen Fundschichten von Salzgitter-Lebenstedt, Norddeutschland. In: B. Ludovici / H. Pöppelmann (eds), Die Tierknochenfunde der mittelpaläolithischen Jägerstation von Salzgitter-Lebenstedt. Forschungen und Berichte des Braunschweigischen Landesmuseums N.F. 1 (Braunschweig 2017) 257-261.

Sutcliffe 1970: A. J. Sutcliffe, Spotted hyena: crusher, gnawer, digester and collector of bones. Nature 227, 1970, 1110-1113.

Vigne et al. 2012: J.-D. Vigne / F. Briois / A. Zazzo / G. Willcox / T. Cucchi / S. Thiébault / I. Carrère / Y. Franel / R. Touquet / C. Martin / C. Moreau / C. Comby / J. Guilaine, First wave of cultivators spread to Cyprus at least 10,600 y ago. Proceedings of the National Academy of Sciences of the United States of America 109/22, 2012, 8445-8449.

Vinnicombe 1976: P. Vinnicombe, People of the Eland: Rock Paintings of the Drakensberg Bushmen as a Reflection of their Life and Thought (Pietermaritzburg 1976).

Wehrberger 2013: K. Wehrberger (ed.), Die Rückkehr des Löwenmenschen. Geschichte, Mythos, Magie [exhibition catalogue Ulm] (Ostfildern 2013).

West/Packer 2002: P. M. West / C. Packer, Sexual Selection, Temperature, and the Lion's Mane. Science 297/5585, 2002, 1339-1343.

Wojtal et al. 2020: P. Woijtal / J. Svoboda / M. Roblíčková / J. Wilczyński, Carnivores in the everyday life of Gravettian hunters-gatherers in Central Europe. Journal of Anthropological Archaeology 59, 2020, 101171. DOI: 10.1016/j.jaa.2020.101171.

Yamaguchi et al. 2004: N. Yamaguchi / A. Cooper / L. Werdelin / W. McDonald, Evolution of the mane and group-living in the lion (*Panthera leo*): a review. Journal of Zoology 263/4, 2004, 329-342.

Zeeb-Lanz/Reymann 2019: A. Zeeb-Lanz / A. Reymann, Löwenmenschen und Schamanen – Magie in der Vorgeschichte. Archäologie in Deutschland: Sonderheft 16 (Darmstadt 2019).

Zeuner 1958: F. E. Zeuner, Dog and Cat in the Neolithic of Jericho. Palestine Exploration Quarterly 90, 1958, 52-55.

Züchner 1996: C. Züchner, The Chauvet Cave – Radiocarbon versus Archaeology. International Newsletter on Rock Art 13, 1996, 25-27.

SUMMARY / ZUSAMMENFASSUNG

Big Cats and Humans – Some Remarks on the Interaction with Big Cats in Prehistoric Art

Using the excavation site in Schöningen and a few other middle Pleistocene sites we can observe the use of carnivore bones as raw material used in constructing tools. It is not clear whether or not the bones were selected purely through their suitability as a raw material or if there was a possible link to the types of species chosen. From Middle Paleolithic excavation sites we have several bones showing signs of having been worked with cut marks etc., this possibly bears witness to an active tendency to specifically hunt carnivores.

We must clearly pay thanks to excavation sites dating to the Upper Palaeolithic in providing clear evidence for the active hunting of carnivores such as the cave lion. In the Aurignacien pendants/necklaces made using teeth, and early works of art area points towards the special role of carnivores within the Human race. Next to the ivory sculptures recovered from the Aurignacien layers in caves faunal in the Swabian Alb/Jura, a special mention must be made of the 30,000 year old paintings from the Grotte Chauvet in southeastern France. Carnivores, using the example »Cave lions«, are often depicted as being a very strong species often portrayed in the context of a »pride« or an active hunting group. An aggressive tendency is not apparent. It is unclear whether or not the obvious closeness of a »lion mane«, reflects the actual appearance of Palaeolithic lions in general. Carnivores, in the context of Ice Age art are, in general a seldom repeated motive, but the position in which they are depicted in the caves and their context, show the special reference afforded them by humans, and that this reverence probably began in the Lower and Middle Palaeolithic.

Großkatzen und Menschen – einige Anmerkungen zur Interaktion mit Großkatzen in der prähistorischen Kunst

An der Fundstelle Schöningen sowie an wenigen weiteren mittelpleistozänen Fundstellen beobachten wir die Nutzung von karnivoren Resten als Rohmaterial für Werkzeuge. Es ist unklar, ob die Knochen allein wegen ihrer Rohmaterialeigenschaften gewählt wurden, oder die Spezies auch von Bedeutung war. Aus mittelpaläolithischen Fundstellen liegen vereinzelt Knochen mit Schnittspuren vor, die möglicherweise die aktive Jagd auf Karnivoren bezeugen. Klare Belege für die aktive Jagd auf Karnivoren wie Höhlenlöwe verdanken wir wiederholt jungpaläolithischen Fundstellen. Im Aurignacien lassen Tierzahnanhänger und frühe Kunstzeugnisse auf eine besondere Rolle von Karnivoren für den Menschen schließen. Neben den Elfenbeinskulpturen aus den Aurignacien-Schichten in Höhlen der Schwäbischen Alb sind hier vor allem die etwa 30000 Jahre alten Malereien aus der Grotte Chauvet in Südost-Frankreich zu erwähnen. Die Karnivoren bzw. Höhlenlöwen sind wiederholt als starke Tiere und wohl auch als aktive Jagdgruppe dargestellt; ein aggressives Verhalten ist jedoch nicht abzulesen. Es bleibt unklar, ob das Fehlen der Darstellung von Mähnen auf das tatsächliche Erscheinungsbild der Löwen zurückgeht. Karnivoren sind insgesamt in der Eiszeitkunst ein eher seltenes Motiv, aber u.a. die Position der Darstellungen in Höhlen sowie ihr Kontext weisen auf eine besondere Rolle dieser Tiere für den Menschen hin, die wahrscheinlich im Alt- und Mittelpaläolithikum ihren Ursprung findet.

NICOLAS ZENZEN

THE LION IN ANCIENT ART – THE STORY OF A STRIKING SUCCESS

The lion is distinguished as the largest and most impressive predatory cat in history[1]. This role is especially reflected in arts, where lions are a common theme. In our everyday life, pictures of lions appear everywhere: in coats of arms, on buildings, monuments, fountains, in gardens or even in Chinese restaurants. This last example is meant to emphasize that we are not dealing with a phenomenon limited to the scope of Europe, but that depictions of lions are as widespread as East Asia.

Tracing back the tradition of all these depictions to their origins we end up in the ancient Near East. There we can find the distinctive scenes and contexts in which lions are depicted again and again, even to this day. As a result, we are dealing with about six millennia of the lion's pictorial history, extending over large regions of the world. The consistency and the narrow range of meanings applied to the lion's representations are remarkable. Notably, it does not seem to matter whether or not lions actually live in the respective environment of a visual culture. Simply the knowledge of the existence of the lion seems to be enough for its extraordinary success as a reoccurring subject in pictorial art across time and cultures.

Hardly anyone can escape the fascination of an actual lion. On the one hand, it evokes fear and terror, and on the other it demands admiration and respect. This is based on the absence of natural enemies, its powerful and elegant appearance, and the charisma of majestic calmness in combination with the explosive violence of a hunt. Appropriately, lion representations contain a well-defined range of meanings, independent of the particular cultural and historical context.

LION ATTACK

The confrontation of one or more lions with prey animals is a common theme in pictorial art. All phases of the attack are shown, from the ambush to the devouring of the victim. Compositionally, all imaginable possibilities to implement the attack are used, as well as a large variety of prey ranging from wild to domesticated animals, and from buffalo to hare.

The most popular of prey animals is undoubtedly the bull, as depicted in an early Sumerian limestone jug from Uruk (**fig. 1**). A standing bull, attacked by a

Fig. 1 Lion attacking a bull. Stone vase with relief decoration from Warka, Iraq. Early 3rd mill. BC. Baghdad, Iraq Museum Inv. 19169 (currently lost). – (Bildarchiv Foto Marburg, photo M. Hirmer; https://www.bildindex.de/document/obj20961337?medium=fm1881584).

1 The tiger might be seen as physically impressive, however, due to a more limited geographical distribution, it is less common in art. To look at its depictions more closely would be exciting, however, it will not be explored further in this article.

In: Nicholas J. Conard · Henning Hassmann · Kurt Felix Hillgruber · Jordi Serangeli · Thomas Terberger (eds), The *Homotherium* Finds from Schöningen 13II-4. Man and Big Cats of the Ice Age. Forschungen zur Urgeschichte aus dem Tagebau von Schöningen 4 (Mainz 2022). DOI: https://doi.org/10.11588/propylaeum.1006.c13526

Fig. 2 Lion attacking a bull. Relief at the staircase of the »Apadana« in Persepolis, about 500 BC. – (Photo N. Zenzen).

Fig. 3 Lioness devouring a calf. Pedimental sculpture from the Athenian acropolis, Poros stone. Early 6th c. BC. Athens, Acropolis Museum Inv. 4. – (Photo © Ad Meskens, Athens Acropolis Museum Hekatompedos Lioness pediment, CC BY-SA 4.0).

lion, is visible in the relief depictions on both sides. The beast has climbed onto the bull's back, striking its front paws into its abdomen. A very similar composition on a relief from Persepolis, depicted on the front of the Achaimenid King of Kings' palace, is over two thousand years younger (**fig. 2**).

Depictions of lion attacks were not only common in the ancient Near East, but also widespread in the Mediterranean region. An example from Greek culture is an adorned gable of a temple on the Acropolis in Athens dating to the early 6th century BC (**fig. 3**). Here a calf lies already lifeless on the ground, while the huge cat of prey, which despite its mane can be recognized as a lioness by the presence of her teats, crouches over it staring at the observer threateningly before feasting on its hind leg. The male counterpart was probably depicted symmetrically to the lioness in the other, not preserved half of the gable.

The theme of the lion attack was depicted with different kinds of prey animals, in a wide range of iconographic compositions, in all thinkable formats and genres, from miniature seal images to monumental free-standing sculpture, as well as paintings. It was common in societies for which the threat of lions is a realistic experience. The lion with all its physical prowess is characterized as a powerful ruler over nature. This predestines the lion – according to its title as »king of animals« – to be used as an analogy for human rulers and powerful deities. Their violence towards enemies is thereby intangibly demonstrated to be impressive. However, at the same time, these images reveal the antagonism between the orderly, beneficial culture of man and the irregular, threatening power of wild nature. From this perspective, the lion – in seeming contradiction to what has just been described – symbolizes the world opposed to man, and in doing so is the antagonist of the representatives of human society, i. e. of gods, heroes and rulers.

LION HUNT

Following the interpretation of the lion as an opponent of rulers, the meaning of the lion hunt as a pictorial theme is apparent. Rulers appear as hunters, for example as early as the Chalcolithic stela from Uruk, in which a bearded man attacks a lion with a spear in the upper register and fires arrows at several of the predatory cats in the lower register (**fig. 4**).

Particularly impressive implementations of that theme are found in the series of reliefs from the Neo-Assyrian palaces, for example of King Assurnasirpal II. (883-859 BC) at Nimrud (**fig. 5**). The ruler in his chariot is shooting arrows at a lion, which, although already wounded several times, jumps aggressively towards the chariot. Another lion, apparently mortally wounded, lies below the hooves of the chariot horses. Two men who beat on their shield with sticks in order to drive the lion towards the king underline the realistic character of the representation.

By presenting himself as a fearless and superior lion hunter, the ruler demonstrates how he fulfils one of his central tasks, namely to protect his subjects from threats and to secure man's dominant position in the world. Particularly in autocratic systems, this was an excellent opportunity for the monarchs to put themselves in the spotlight. However, the depiction as a lion hunter was only credible proof of power if the hunt actually took place. As a result, the corresponding iconography is only found with rulers who were able to hunt lions in reality.

In this context it is interesting to take a look at the societies of the Aegean. During the Bronze Age these societies were already less autocratic, and were led by warlike elites with egalitarian principles. The theme of the lion hunt is also present in Aegean art, however not one single individual appears as an opponent of the lion, but rather a group of hunters. A wonderful example is the bronze dagger with a metal inlay depiction from one of the Mycenae shaft tombs (**fig. 6**). Five hunters equipped with large shields, spears, bows and arrows, attack three lions, two of whom flee. Yet, the largest of the three lions turns against the attackers and throws the foremost of them to the ground. Thus, the danger of the hunt and the vulnerability of the hunters are dramatically portrayed. The message is obvious: only within a community people are able to persist amidst the world's dangers.

Fig. 4 Lion hunt. Basalt stele from Warka, Iraq. Late 4th mill. BC. Baghdad, Iraq Museum Inv. 23477. – (alamy.de, ID: WH98P1).

Fig. 5 The Assyrian king hunting lions. Relief from the palace of Ashurnasirpal II. Gypsum, 865-860 BC. London, British Museum Inv. 124534. – (akg-images, AKG1095720).

Fig. 6 Lion hunt, bronze dagger with inlaid work from shaft grave 5 in Mycenae, 17th c. BC, Athen, National Archaeological Museum Inv. 394. – (Photo Zdeněk Kratochvíl, CC BY-SA 3.0).

Fig. 7 Lion hunt. Section of the »Alexander sarcophagus« from the royal necropolis near Sidon, Lebanon. Marble. Late 4th c. BC. Istanbul, Archaeological Museum Inv. 370. – (Photo N. Zenzen).

Fig. 8 Heracles fighting against the Nemean lion. Vase painting on an attic amphora from Vulci, Italy. about 530 BC. Munich, Staatliche Antikensammlungen Inv. 1395. – (© Landeshauptstadt Hannover, Museum August Kestner, photo Ch. Tepper).

In the relief of the so-called Alexander sarcophagus from Phoenician Sidon, Greeks and Orientals, recognized by their respective clothing styles, together prey on a large lion (**fig. 7**). They fight on horseback and on foot, wielding an array of weapons against the cat of prey, which has already struck the chest of one of the horses. Within its particular historical context the successful effort of the lion hunt demonstrates the strength of the elites in Alexander's empire, created by the unification of East and West. In fact the Macedon king presented himself as a fearless lion hunter. Thus, he followed the tradition of the oriental rulers on the one hand, and is personifying the new Heracles from a Greek perspective on the other.

LION'S SUBJUGATION

In the Greek world, before Alexander the Great, only mythical heroes – specifically Heracles – had the right to defeat a lion on their own. The fight of Heracles against the lion of Nemea is one of the most popular subjects of Athenian vase painters, as in the example on an amphora of the 6th century BC found in Etruscan Vulci (**fig. 8**).

However, the theme is much older, exemplified by a cylinder seal from Mesopotamia, which was made about 1800 years earlier (**fig. 9**). The characteristic pose of the erect lion, pressing its rear paw against the

Fig. 9 Heroes fighting against lions. Akkadian cylinder seal. Serpentine. 24th-23rd c. BC. London, British Museum Inv. 104489. – (akg-images, AKG429560).

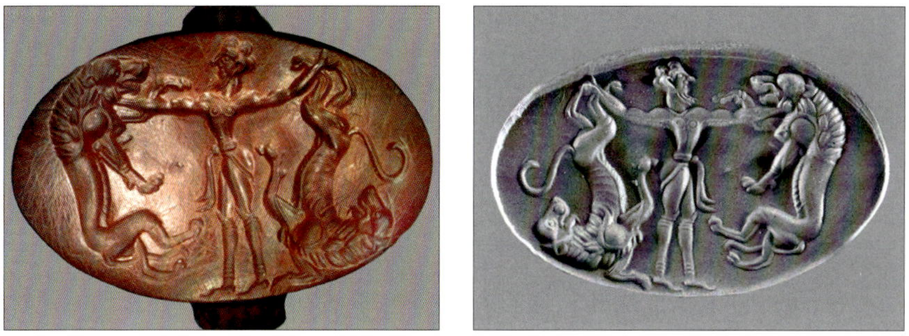

Fig. 10 »Mistress of animals«. Lentoid seal, carnelian, and impression. From a chamber tomb near Mycenae. 16th c. BC, Athens, National Archaeological Museum Inv. 6442 ε. – (Corpus der minoischen und mykenischen Siegel am Institut für Klassische Archäologie der Universität Heidelberg I 144).

leg of his human opponent, is hardly based on observation in real life, but rather an artistic invention, suggesting a continuous iconographic tradition over the centuries.

In contrast to the hunting scenes, these representations only refer to a fictitious reality. Though, the subjugation of a lion by a human in the form of hand-to-hand combat is still the version closest to common experience.

Even more imaginary, though impressive, is the scene of the so-called »Master of Animals«, who holds a full-grown lion in each hand. The figures depicted doing this are extremely strong and powerful. Therefore, they should be considered as superhuman beings.

Even more popular than the male master, was the theme of the »Mistress of Animals«. The female figures in this position usually do not have to use any physical force to keep the lions in check. Depicted on a carnelian seal from a grave near Mycenae, the central female figure with her oversize crown is flanked by cats of prey standing on their hind legs as if they were trained (**fig. 10**). It is obviously a goddess, under whose power even the wild animals are subordinate.

However, Greek and Ancient Near Eastern goddesses can definitely be ruthless and violent, when acting against enemies. The lions subordinated to their divine power may demonstrate this quality. This becomes particularly clear in the widespread theme of the goddess riding a lion. The Mesopotamian goddess Ishtar usually stands upright on a lions back, as seen in an Assyrian relief stele from Til Barsip in present-day Iraq (**fig. 11**). The armament exemplifies the fighting nature of the goddess, and, as she keeps the lion on a lead, the subordination of the animal is emphasized. Greek goddesses are more likely to go to battle sitting on a lion. An example of this is can be seen from the mother goddess Cybele (or Rhea) at the Pergamon Altar in the battle against the giants (**fig. 12**). The lion aids the deity not only as a means of transport, but also as a combatant. It fights together with the gods against the giants that threaten to overthrow the existing world order. Thus the lion's strength is used in the service of the gods and human civilization.

Fig. 11 Ishtar on the lion. Stele from Til Barsip, Iraq. Red breccia. 7th c. BC. Paris, Musée de Louvre Inv. AO 11503. – (akg-images, AKG329897).

Fig. 12 The goddess Kybele or Rhea riding on a lion in the battle against the giants. From the great frieze (south side) of the Pergamon Altar. Marble. 1st half of the 2nd c. BC. Berlin State Museums. – (akg-images, AKG1914652).

Fig. 13 Grave lion from the lion tomb at Miletus, Turkey. Marble. Mid-6th c. BC. Berlin State Museums Inv. SK 1790. – (BPK Bildagentur, 000188887).

GUARDIAN LIONS

A similar idea of lions as serving humans can be observed in the depiction of lions as guardian figures. Such sculptures often flank entrances to shrines, palaces, or cities in pairs, a practice that can be traced back to the end of the 4th millennium BC in Mesopotamia. In most cases these lions open their mouths aggressively and look awe-inspiringly at those entering the gate.

An example from Bronze Age Anatolia are the colossal lions at one of the gates to the Hittite capital Hattusa. The famous Lion Gate of Mycenae, the main entrance into the citadel is from the same period. There, the two cats of prey are perched in the tympanum field above the entrance. By supporting their forelegs on a podium carrying a column, they stand symmetrically upright in a majestic pose. The heads of the lions were fitted separately and therefore did not survive, however it is undoubted that they looked frontally at the person entering and thus demonstrated their vigilance.

Regardless of what the figures at the gate look like, their function remains clear: The lions are in the service of the human sphere and protect the central institutions of a society. Although real lions hardly behave in this way, the image is thoroughly convincing. In iconographic terms, the function as guardian is often underscored by staring eyes that directly fix the entering person, and by their mouths that are threateningly wide open.

Guardian figures at entrances are not very common in classical antiquity, whereas lion statues are often found at graves. There they can have a wide variety of poses. One example is a marble lion from an archaic grave in Miletus (**fig. 13**). Adequate to the solemn atmosphere of a tomb, he lies calmly, yet vigilantly.

The figures of guardian lions, however, demand much more than awe from the viewer. They also make a statement about the person they are guarding. Only people strong enough to subjugate the predator are worthy of being guarded by a lion. Therefore, the guardian lion is used to signify an outstanding personality in those who they are watching over, regardless of whether that person is deceased, living, a ruler, or a deity.

LIONS AS SYMBOLS OF POWER

The connection between power and lions run through all of the demonstrated themes. The relentless predator serves as a metaphor for the fighting qualities of the powerful. On the other side rulers demonstrate that they are capable of protecting their subjects from danger through images of hunting and fighting against lions. In the same way lions emphasize the greatness of men's power by being depicted as guardians of human institutions. Therefore it is hardly surprising that the image of the lion as a symbol of power and dominion takes on a life of its own. In royal contexts, for example, there are consistently images of lions. Their meaning becomes clear only by the presence of the real person of the ruler. They underline his majesty and can even represent his power in his absence.

The fact that the lion can be both the opponent and the parable of the ruler only seems like a paradox, because actually one is the consequence of the other. The example of Heracles is illustrates this connection: After he has defeated the lion, Heracles skins the cat and wears its fur from then on as a trophy. Through this act, Heracles slips into the role of the predator and ingests its strength. Thus, the inherent ambivalence, which hero and lion have in common, becomes clear: the admired vigor inevitably goes hand in hand with unscrupulous violence. This also applies to rulers or deities in general, because they solve the threat posed to culture (embodied in the form of a lion) by slaying it, but in the face of other adversaries they themselves become raging predators. This ambiguity, which is inherent in the image of the lion, has certainly played a large part in its continuing popularity on the basis of ancient tradition. This is not limited to European pictorial art, but also spreads from the Near East throughout the entire Asian continent and beyond. Numerous examples of the characteristic themes mentioned here can be found, even as far as Sri Lanka, Indonesia and Japan.

Particularly in cultural contexts in which there are no real lions, the significance of the lion as a symbol of domination tends to be superimposed onto other aspects. Positive connotations, such as courage, pride, and majesty prevail against the more negative ones, such as cruelty, unpredictability, and destructiveness. Nevertheless, it seems crucial that there is an awareness about the inherent danger in the lion as a predator, so that the positive connotations can have their effect. It is this multi-layered, yet complex system of meaning, combined with the impressive and artistically graceful figure that has made the lion into such a popular and enduring subject of human pictorial art.

SUMMARY / ZUSAMMENFASSUNG

The Lion in Ancient Art – The Story of a Striking Success

The lion is the most frequently depicted animal in global art. The phenomenon can be traced back to about 5,000 years ago to the art of the Ancient Near East. Here, a system of meanings around the image of the lion was developed. This became the base for lion depictions all over the world. There is a well-defined range of sceneries, in which the lion is depicted, and they all follow a coherent range of meanings. The most common themes are treated here: the lion attack, the lion hunt, the lion's subjugation, the guardian lion, and the lion as symbol of power. On the one hand, the lion represents the wild nature and its threatening aspects, and is thus an opponent to human culture. On the other hand, the lion acts as an analogy for human strength and power, becoming a defender of culture. This change between two seemingly contradictory meanings makes the lion such an attractive subject of pictorial art.

Der Löwe in der antiken Kunst – die Geschichte eines bemerkenswerten Erfolgs

Der Löwe ist in der globalen Kunstgeschichte das am häufigsten dargestellte Tier. Das Phänomen lässt sich bis in die Kunst des Alten Orients vor etwa 5000 Jahren zurückverfolgen. Hier wurde ein System von Bedeutungen rund um das Bild des Löwen entwickelt. Dieses bildet die Grundlage für Löwendarstellungen in aller Welt. Sie folgen einem klar umrissenen Repertoire an Bildtypen und umfassen ein schlüssiges Spektrum an Bedeutungen. Die wichtigsten Motive sind: der Löwenüberfall, die Löwenjagd, die Unterwerfung des Löwen, der Wächterlöwe und der Löwe als Machtsymbol. Einerseits steht der Löwe für die wilde Natur und deren bedrohliche Wirkung, wodurch er zum Gegenbild der menschlichen Kultur wird. Andererseits tritt er als Gleichnis für menschliche Stärke und Macht und somit als Verteidiger der Kultur in Erscheinung. Dieses Wechseln zwischen zwei scheinbar widersprüchlichen Bedeutungen macht den Reiz des Löwen als Motiv der Kunst aus.

MODERN CATS – CARNIVORES

ULRICH SCHMÖLCKE · ELENA A. NIKULINA

HUMANS AND THEIR RELATIONSHIP TO LARGE CARNIVORES IN CENTRAL EUROPE FROM FORAGERS TO MODERN TIMES: A SURVEY

Through the whole history of mankind, the relationship between humans and large carnivores was a mixture of fascination and dread. They were admired and feared, cosseted and killed (Kruuk 2002). Even if such emotions are not tangible in archaeological remains, there are several possibilities to prove the humans view on these animals, e.g. the presence (and also absence) of skeletons of carnivores or traces on their bones made by man. The scientific discipline taking a holistic view on the interactions between humans and animals in the past is named Social Archaeozoology or Social Zooarchaeology (Russell 2012). It does not only focus on archaeological data such as remains of animals, but includes also results of ethnographic studies in the investigations as well as analyses of depictions and of written sources. An understanding of all facets of relationships between animals and humans needs the cooperation of many different scientific disciplines, and consequently a comprehensive volume about this topic includes perspectives from zoology, ethnology, neuropsychology, developmental psychology, human medicine, sociology, jurisprudence, cultural sciences, philosophy, theology, and history (Otterstedt/Rosenberger 2012). An approach similar to the mentioned volume would also be necessary to tell a complete story about the relation between humans and large carnivores in prehistoric and historic times. Combined together, these sources provide detailed information about the status, prestige and reputation of carnivores and show that they have played numerous roles in past human societies: companions, helpers, diet, sacrificial victims, scary manifestations of »the evil«, objects of taboos and more (Russell 2012).

Especially the interactions between humans and carnivores – this name includes canids, felids, hyaenids, and ursids – during the Ice Age are in focus of scientific research since decades (summarised in Rosell et al. 2012). Direct records of carnivores such as bones or indirect evidence for their presence like remains of their prey are known from many campsites of Ice Age people. In most cases this close spatial contact was possible because humans and carnivores alternated in using the same sheltered caves or dens (Turner 1992). Potentially, contacts of humans and large carnivores often originated in visits of carrion-feeding predators in shelters where humans had stored their meat (Rosell/Blasco 2009).

Until the Ice Age a kind of coevolution between carnivores and humans can be observed (Brantingham 1998; Stiner 2002; Schleidt/Shalter 2003). At the end of such a process at least between wolves and humans a coexistence developed. Probably via the step of commensalism this coexistence resulted in the first domestication of an animal species in human history.

Most of the Ice Age carnivores are extinct today: sabertooth cat (*Smilodon fatalis*), cave lion (*Panthera spelaea*) and cave bear (*Ursus spelaeus*) disappeared already before the Ice Age ended (Stuart/Lister 2011; Diedrich 2014; Stuart 2015). The spotted hyaena (*Crocuta crocuta*), widely distributed in Africa, was also part of the European fauna during the Ice Age, but extinct there 30000 years ago at the latest (Stuart/Lister 2014).

The present paper focuses on the last 15000 years, i.e. the Holocene, the time since the ending of the last Ice Age. The geographical emphasis will be Central Europe and in some parts also Scandinavia. In this area

In: Nicholas J. Conard · Henning Hassmann · Kurt Felix Hillgruber · Jordi Serangeli · Thomas Terberger (eds), The *Homotherium* Finds from Schöningen 13II-4. Man and Big Cats of the Ice Age. Forschungen zur Urgeschichte aus dem Tagebau von Schöningen 4 (Mainz 2022). DOI: https://doi.org/10.11588/propylaeum.1006.c13527

during the Holocene only four larger carnivore mammals shared the same habitat with humans: brown bear (*Ursus arctos*), wolf (*Canis lupus*), lynx (*Lynx lynx*), and wildcat (*Felis silvestris*). Others, such as leopards (*Panthera pardus*), tigers (*Panthera tigris*) and lions (*Panthera leo*) also reached Europe, but did not occur in areas north of the Balkans and the Ukraine (Sommer/Benecke 2006). Consequently, the relationships between those large cats and humans will be excluded, particularly because several overviews about this topic are available (Yannouli 2003; Thomas 2004; Bartosiewicz 2009; Thomas 2014).
Additionally to bear, wolf, lynx and wildcat two more animals are included into the present study: the domesticated forms of wolf and wildcat. The questions for the ancestors of domestic cats and dogs and for the process of domestication have been clarified by intensive research of anatomists and archaeozoologists (Bökönyi 1974; Herre/Röhrs 1990; Hemmer 1990; Benecke 1994; Clutton-Brock 1999), and their results were nowadays complemented by analysis of ancient DNA in bone remains and by population genetics. In the case of these two animals we briefly summarise the domestication process and then describe the ways of utilising dog and cat in the past. The utilisation of dogs and cats but also their reputation have changed over times and cultures, and this circumstance offers quite deep insights into the relationship and co-existence of man and wild or tamed carnivores.
Several aspects in the present article origin from research of the research focus »History of Hunting and Fishing« at the Centre of Baltic and Scandinavian Archaeology (ZBSA) located in Schleswig/D. This focus includes research projects covering mainly the area between the northern Atlantic and the Ural Mountains and the periods from Late Palaeolithic to the Middle Ages. The ZBSA projects include and combine a large variety of methods used both by arts and sciences (Grimm/Schmölcke 2013; Grimm et al. 2014).
The idea of the present paper is rather to give a short overview about the state of research, complemented by results of recent investigations. Primary and secondary written and pictorial sources, but also finds of weapons and bones are so numerous that their evaluation must be the task of comprehensive volumes (Paul 1981; Kruuk 2002; Fascione/Delach/Smith 2002; Bieder 2005; Kalb 2007; Faure/Kitchener 2009; Brunner 2010; Kindler 2012; Turner/Bateson 2014). Especially considerable is the literature about human/dog relationships, and this field of research in particular can only be touched briefly here (for further information see Crockford 2010 and Morey 2010).

RELATIONS OF FORAGERS AND FIRST FARMERS TO LARGE CARNIVORES

During the Mesolithic, which stretches by definition the cultures and time between the end of the Ice Age and the beginning of farming, human diet was based on hunting, fishing, and gathering, similar to the preceding Ice Age. In the successively forested Central European landscape people lived in temporary settlements and mobile dwellings, probably different kinds of tents (Wenzel 2009). One of the archaeologically best investigated areas of the Central European Mesolithic is the region around the Mecklenburg Bay of the Baltic Sea in northern Germany. Between 1995 and 2005 a series of settlements uncovering hundreds of well-preserved animal remains have been excavated there (Hartz et al. 2014). The *c.* 1500 mammal bones, 750 from Mesolithic and Neolithic each, are not only a suitable basis for reflections about the relationship between humans and carnivores during the Mesolithic, but also for the development of Stone Age hunting strategies in general (Hartz/Schmölcke 2013). Such analysis shows that in the region around the Mecklenburg Bay the people from the Mesolithic hunter-fisher-gatherer cultures mainly exploited red deer (*Cervus elaphus*), wild boar (*Sus scrofa*), aurochs (*Bos primigenius*) and roe deer (*Capreolus capreolus*); from these species derive more than 90 % of animal bones found at human settlements. Apart from such game species also remains of fur-bearing species such as beaver (*Castor fiber*), otter (*Lutra lutra*) and other mustelids are recorded, and sometimes also sealing played a role. But the most relevant aspect for the present paper is that from terrestrial carnivores not a single remain larger than from red fox (*Vulpes vulpes*) was found.

This regional example of animal exploitation during the Mesolithic can be generalised, since remains of large carnivores are commonly exceptional in bone assemblages from European human settlements of the very period (Charles 1997). Although there is no doubt that wolf, brown bear and wildcat were distributed over all parts of Central Europe in that time, their remains are only scarcely found (Sommer/Benecke 2006). It has to be concluded that these species were no regular hunting prey for Mesolithic people. Even the fact that artificially perforated teeth from these carnivore species have been found repeatedly at Mesolithic sites in Central Europe is not contradictory. It just documents the use of teeth of large carnivores as an application of clothes, but such a practice is not unique to teeth of them but also known for teeth of other species (Gramsch 2014). Impressive and significant examples provides the Neolithic cemetery of Ostorf (city of Schwerin/D), where in several human graves necklaces have been found combining hundreds of single teeth of bear and wildcat, but also of smaller carnivores such as badger, otter, pine marten, red fox, and dog, as well as of red deer (Lehmkuhl 2010).

To summarise these data, in Central European hunter-gatherer societies carnivores probably played neither a role as a hunting trophy nor were they considered as food competitors. Potentially, during the Mesolithic time humans and large carnivores avoided direct contact to each other. Alternative explanations for the scarcity of records could be found in the small population density of predators in contrast to herbivores, or hunting carnivores was too risky in comparison with the output, or there were spiritual reasons forbidding bear and wolf hunting.

In the 4th millennium BC with the beginning of the Neolithic, hunter-fisher-gatherer cultures in whole Central Europe were replaced step by step by cultures of farmers. After a time of transition, the new human diet based first and foremost on agriculture and husbandry with cattle, pigs, sheep and goats (Lüning 2000; Feeser et al. 2012). In our focus region around the Mecklenburg Bay besides Mesolithic camps also several Early Neolithic sites have been excavated (Hartz et al. 2014), and the analysis of the mammal remains unearthed there proves a mixed economy (Hartz/Schmölcke 2013): Although the subsistence of the people based mainly on domestic animals now, hunting was still an important part in the life of the first farming communities, and remains of wild mammals still account for 38 % of the bones. 89 % of the bones from wild animal species derive from red deer, again followed by roe deer, wild boar and aurochs. The remaining 11 % come from a long list of animals including brown bear as well as wolf and wildcat. In other parts of Central Europe also lynx can be added to this list of species occasionally hunted by the Stone Age farmers (Sommer/Benecke 2006). In contrast to the previous forager cultures, for the first farmers large carnivores seemingly had become a relatively common object of hunting. This result, which has supra-regional validity (Steffens 2005), may indicate a change in the overall tradition and motivation of hunting. Thanks to the presence and breed of livestock, hunting was not the central dietary source any longer, and as a consequence new, alternative hunting motives and reasons developed: religious, symbolic and ideological as objects of prestige (Boyle 2006; Hartz/Schmölcke 2013). The chase of carnivores was in those days probably mainly undertaken in order to protect the livestock. However, in spite of more intensive hunting of larger carnivorous species, the human pressure had no influence on their population size and distribution. Aside from isolated populations on islands, there is no evidence that the first farmers reduced the natural number of these animal species purposeful and effectively (Butzeck/Stubbe/Piechocki 1988; Sommer/Benecke 2005).

Fig. 1 Brown bears are the largest terrestrial carnivores in Holocene Europe. – (Photo Francis C. Franklin, European Brown Bear, CC BY-SA 3.0).

HUMANS AND BROWN BEAR

Prior to times of cultures with written sources or depictions it is difficult to reconstruct the perception of the largest Central European predator, the brown bear (**fig. 1**), and the general relationships between bear and man. It is only after such archaeological and historical sources complement the bone assemblages our knowledge of the human/bear-relationship increases (see Schmölcke/Groß/Nikulina 2017 for many examples). Generally, the typically very small number of brown bear remains in European archaeological bone assemblages in the whole period between Stone Age and medieval times demonstrates that bears played a minor role as an object of hunting over millennia. This observation is in contrast to the value of brown bear meat, which contains more calories per kilogram than all other terrestrial mammals available for the people (U.S. Department of Agriculture 2016). Probably, the minor relevance of bear meat is connected to their low population density, but surely also to the fact that hunting bears is widely known as a highly dangerous and risky undertaking. The risk in meeting bears is also mirrored in their name in the Germanic languages, since the English *bear*, the Scandinavian *björn* or the German *Bär* root in the old **bher-* or **bheros*, originally meaning just »brown« (Pokorny 1959). Using such an indirect name is common when the actual name of an object is a taboo. By calling the species »the brown one« the people avoid to call up the potentially dangerous animal (Kluge/von Seebold 2012). A similar use of euphemisms for the bear is known in Slavonic languages, where the name for bear goes back to verbatim »honey eater« (e.g. Russian медведь, Ukrainian ведмідь, Polish *niedźwiedź*; see for instance Černych 1999, 781-782). For different peoples in Siberia the bear is either »the old man« or »the grandfather« (Duerr 2010, 73).

Even if there was no relevant economic interest of Stone Age to Early Iron Age people in bears as a meat resource, in spiritual and religious realms, however, the things might have been very different. Many native people of the north of Europe, Asia, and America admire the brown bear as a kind of mediator between animals and humans until modern times, and for that reason they normally did not kill them. Often it was only allowed to hunt a bear after performing strict and complex rituals before and after the hunt (Duerr 2010, 76ff.; Wamers 2015). Much could be said about bear symbolism, about the meaning and reputation of the species in the past (Schmölcke/Groß/Nikulina 2017). In the present paper only two widely unknown fassets should be mentioned: In several cultures bear hunting became an ideological context in the course of time with high prestige, and bears became »the hero's enemy« (Oehrl 2013). As S. Oehrl has described and analysed in a comprehensive paper about this topic, from late antiquity onward bear hunting was regarded as a particularly heroic deed, demanding great courage and earned the hunter glory. Following Oehrl's detailed explanation the roots of this motif go back to the 2nd century AD, the reign of Roman Emperor Hadrian. The so-called Arch of Constantine in Rome, originally dedicated to Hadrian, for instance, shows the depiction of a bear hunt among others, and for this Emperor the symbolic significance of his personal success in hunting bears was so important that he even coined money with the bear as a motif. From Hadrian's time to the Middle Ages, brown bears became prestigious hunting trophies, insignia of heroes with an extraordinary deed of courage.

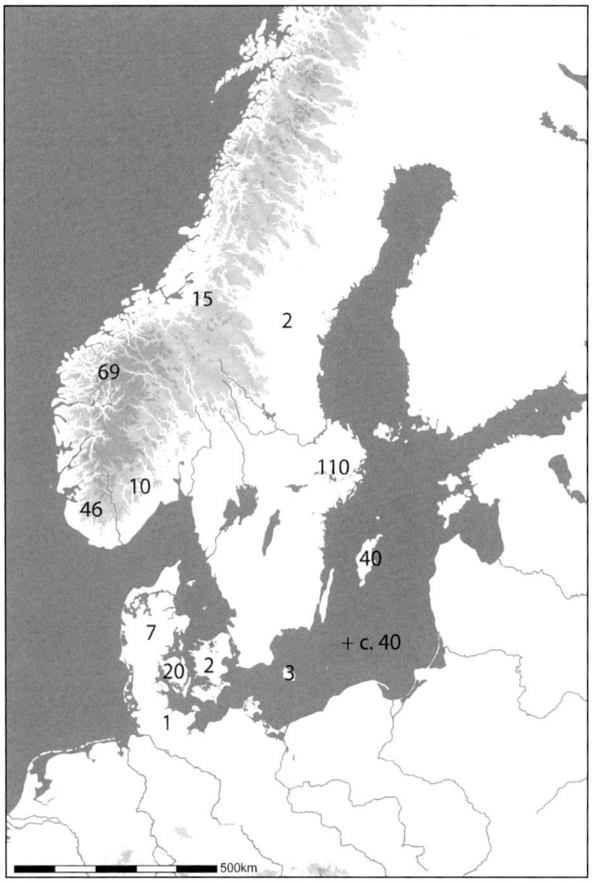

Fig. 2 Distribution map of human graves with skins or claws of bear. All graves dated to the 1st millennium AD. – (Changed after Grimm 2013, figs 2. 5).

Contemporaneously to this development, in northern and eastern Central Europe, but especially in Scandinavia another tradition can be observed, potentially arising independently: from the last centuries BC until the Christianisation in this region (dating regionally different, starting about AD 300 and ending about AD 1000) in human graves remains of bear appeared quite regularly (**fig. 2**). These remains are often claws, which might indicate the former presence of a bearskin under the dead person, in some cases even skins are preserved (Grimm 2013, fig. 1). In total more than 400 burials with bear-related furnishings dated to the 1st millennium AD are known from Scandinavia, and at least 100 in Central Europe, so this tradition became almost a »mass phenomenon« (Grimm 2013). As O. Grimm has demonstrated in his thorough review of this feature, human graves with bear remains have a marked concentration in mid-east Sweden, and in particular their frequent presence at Gotland in the Baltic Sea is remarkable. Since brown bears did not occur on this island, their remains on Gotland reflect regional fur and skin trade (Petré 1980). A second contemporaneous regional concentration of human graves with bear skins was western Norway; in this area the distribution is strongly linked in time and space to changing traditions of the burial rite (inhumation burials versus urns) and favourable preservation conditions. Most of the Norwegian burials belong to persons of high social rank, both men and women. As O. Grimm pointed out, the oldest known graves with bear-skins in Norway as well as in Sweden are those of women. It is remarkable that the richest known graves from the mid of the 1st millennium AD in Scandinavia contained skins, whereas in other cases, only single sometimes perforated claws or bear-tooth have been put in the grave.

The meaning and the background of this burial tradition is still under discussion. Potentially, these graves were burials of bear-fighters, and this suggestion is supported by pictorial sources, in which the bear-fighter motif is quite popular during the 1st millennium AD in the centuries in question (Wamers 2009). However, the frequency of female graves with bear furnishing in particular raise doubts about a strong connection between successful bear hunters and such grave goods (Grimm 2013). An alternative explanation is given by S. Oehrl (2013): It could have been the husband of the dead woman who gave her his most valuable hunting trophy into the grave to demonstrate the rulers of the afterlife that the deceased was the wife of a hero.

After Christianisation the image of bears changed and in the imaginary of the people bears shifted from powerful and respected kings of the woods to redoubtable »beasts« with a close connection to the devil (Molsdorf 1926, 133). Potentially for the purpose of making the devil tamed and conquered, in medieval times the practice of dancing bears spread over Europe (Bieder 2005; Brunner 2010, 139-146). Already during the Middle Ages bears were also kept in special bear-pits as attractions (Hauck 1963; Butzeck/Stubbe/Piechocki 1988). In those times, at least in parts of Central Europe the people were obligated by the authorities to kill every single bear, if possible, and after registration the hunter was allowed to keep the dead bear (Lindner 1940).

The increasing hunting pressure, but also the intensive medieval forest cleaning in the course of the growing human population density fragmented the distribution range of brown bears in Central Europe (Müller 1971; Butzeck/Stubbe/Piechocki 1988; Benecke 2000). Originally occurring throughout Europe, with the exception of some islands, they disappeared in the course of the last millennium from many areas. However, in most parts of Europe it was not until the 19th century that bear populations declined dramatically (Huber/Swenson in: Deinet et al. 2013). Today the total number of brown bears in Europe increased again and amounts about 17 000 animals added by more than 10 000 in the European part of Russia. This recent development is the result of the youngest change in the human/bear-relationship (presented e. g. by Kalb 2007 and especially obvious in Bürglin 2015).

HISTORY OF DOMESTIC DOGS AND HUMAN RELATIONSHIP TO WOLVES

Wolves (*Canis lupus*; **fig. 3**), in former times widespread in the northern hemisphere, were the first domesticated animals in human history. Due to anatomical and ethnological studies at the end of the 20th century there was no doubt any more that this species is the only ancestor of the dog (summarised in Herre/Röhrs 1990; Clutton-Brock 1995). In the meantime phylogenetic studies have verified this result (Vilà et al. 1997; Vilà/Maldonado/Wayne 1999; Wayne/Ostrander 1999). As very intensive research during the last decade shows, the process of domestication took place independently at different places at least in the Near East, in Siberia, China, and Europe between 32 000 and 14 000 years ago (e. g. Germonpré et al. 2009; Larson et al. 2012; Thalmann et al. 2013; Morey 2014; Skoglund et al. 2015; Wang et al. 2016).

One of the oldest dog skeletons known so far has been found associated with a Late Palaeolithic double burial at Bonn-Oberkassel in western Germany; it is approximately 13 500 years old (Nobis 1986; Street 2002). In this case members of a Late Glacial population of *Homo sapiens* have buried a canid next to a human couple. The fact that one of the oldest securely identified dogs is also the oldest known dog burial un-

Fig. 3 For the longest time of the Holocene wolves were widely distributed in most parts of Europe. – (Photo G. Kramer, U.S. Fish and Wildlife Service).

derlines impressively the emotional significance of dogs already in early stages of domestication. Obviously, with the domestication a strong social relationship between dogs and people developed. From that time on dog burials became a worldwide phenomenon until today (Wiedemann 2002), and some archaeological sites contain hundreds or even over one thousand buried dogs (Morey 2006, tab. 1).

Based on these features, but also because of finds of scattered dog remains in settlement contexts it is possible to reconstructed several details about Stone Age dogs: measurements of bones allow for instance the reconstruction of size (Benecke 1987; 1994; Morey 1992), and the diet of the animals is reflected in the stable isotope composition of their remains (Ewersen et al. 2018). However, it is difficult to determine the main functions of the earliest dogs (Horard-Herbin/Tresset/Vigne 2014, tab. 3), since no written or pictorial sources are available from this period of time. Archaeozoological studies can find evidence if dogs were used as food or for fur or as religious symbols and emotional objects, but they cannot answer the question if these dogs were also hunting assistances, herder, guards, labourer or bed-warmer (Manwell/Baker 1984; Russell 2012, 279-296). From Central Europe features and indications for nearly all these functions can be collected, showing a complex picture with dogs as beloved partners as well as regular food resource (Vang Petersen 2013; Ewersen/Ramminger 2013). From our modern Central European point of view the latter

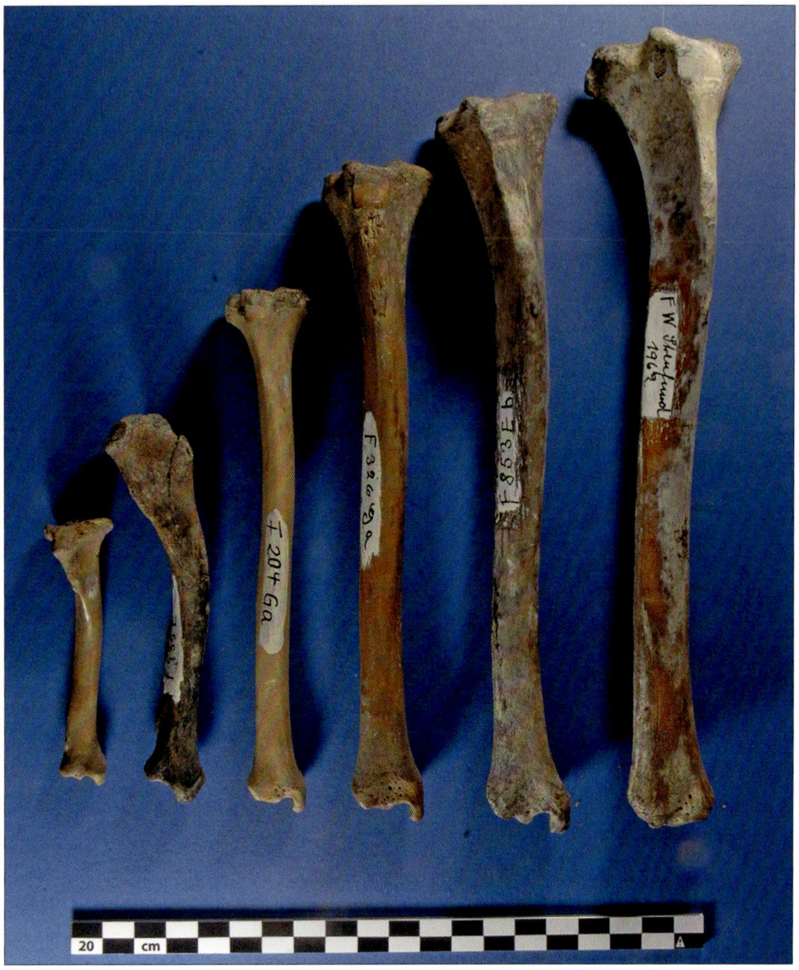

Fig. 4 Influenced by the developed level of Roman dog breeding the variability in size was also high the Germanic area. In this example of tibiae from adult dogs from Feddersen Wierde (Lkr. Cuxhaven/D), the left bone origins from an about 27 cm high dog, the right one from a dog of about 70 cm shoulder height. – (Photo U. Schmölcke).

function of dogs seems to be hard to belief, but factual the purposeful production and use of dog meat was common in some southern German cities until the first half of the 20th century (Geppert 1990). Not to mention living traditions in other parts of the world as in South Korea, where dog meat is after pork, beef, and chicken the fourth most-consumed meat today (Kim 2007; von Chamier 2017).

Without any question, in the Central European Stone Age hunter-gatherer community dogs were not only animals for production, but had also the status as valued companions at the same time. Particularly well-preserved burials of dogs from Mesolithic sites in Denmark and Sweden, but also in the Balkan area emphasise the social importance of these animals and potentially they point to a special role as people's helper also in the afterlife (Vang Petersen 2013). With a view on the presence of dogs in human graves, e. g. at the Swedish site Skateholm (Skåne län/S; Larsson 1990; Fahlander 2008) and the ethnographic record it has been suggested that dogs had a semi-human status at least in some Scandinavian Mesolithic cultures (Tilley 1996, 35). The buried dogs were thus interpreted as substitutions for missed humans, as animal shamans, as symbolic watchdogs or dear members of the household (for detailed discussion see Fahlander 2008; Morey 2006).

In Central Europe and southern Scandinavia, comparisons of the size of carnassial teeth of Mesolithic dogs and wolves have shown that the shoulder height which is about 80 cm in wolves was only 50 cm in mean in dogs (Benecke 1994). This body size was consistent for a long time in many parts of Europe. In Old Egypt, however, and later in ancient Greece and Rome special types of dogs were bred including small pets, dogs similar to modern sighthounds and large, strong and aggressive variances (Peters 1997; Schmölcke

2013). Influenced by the Romans, about 2000 years ago a higher variability in size and stature in dogs developed also in Central Europe. The majority of the Germanic dogs, however, still reached shoulder heights of about 50 cm, but the Roman provinces along Rhine and Danube were an exception. Nearly each excavated settlement in this area derives remains from adult dogs of very short stature, with shoulder heights less than 30 cm, sometimes less than 20 cm (**fig. 4**). This record is in accordance with written sources which also mention such small animals as toy dogs (Peters 1997). In the Roman parts of Central Europe 2000 years ago the phenotypical differentiation in dogs was so distinctive that a pre-stage of selective breeds could be expected (Benecke 1994). With the end of the Roman culture about AD 400 the high variability of dogs came to an end. In the subsequent period the dogs were again less varied in size and stature.

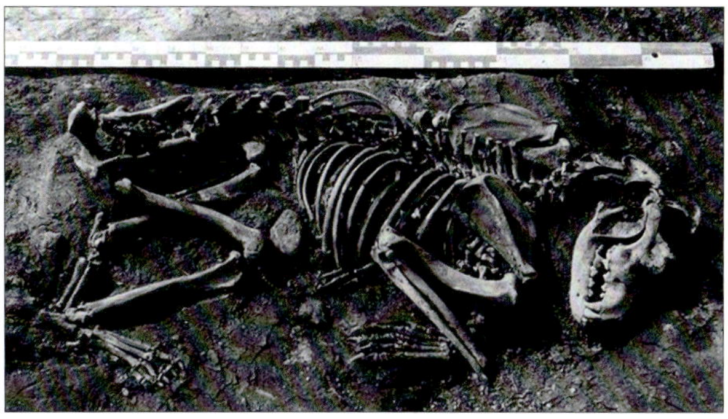

Fig. 5 During the 1st half of the 1st millennium AD burials of dogs in the entrance of houses or between buildings were very common in many parts of Central and Eastern Europe. Feddersen Wierde. – (Photo U. Schmölcke).

In the Middle Age the large majority of dogs had a size similar to the Mesolithic dogs and comparable to modern feral street dogs. In practise of dog breeding the people conducted for a long time a choice for utilisation: out of the whole individual variability within a local dog population, special types have been preferred for hunting, herding, or as watchdogs. Until modern times, there is no evidence for a purposeful development of special dog breeds in Central Europe.

Between 500 BC and AD 400 dogs were buried regularly in large parts of Europe (Makiewicz 2003). Since many similar cultural features at this time can be observed from Denmark over Germany, Poland and Czechia to Romania and Moldavia (Prummel 1992; Makiewicz 2003), one can conclude that the tradition to bury dogs was known in the whole area east of the Roman Empire. In nearly all cases these dogs were deposited in connection with buildings or between houses, anyhow inside the human settlements (**fig. 5**). Therefore, the dog's symbolic role as guards of the houses is obvious (**fig. 6**). However, interestingly the age of such building sacrifice dogs ranges from young puppies to very old animals, so it was not the real power or watch-

Fig. 6 Detail of the dog from **fig. 5**. The forehead of the dog is destroyed, potentially by an intentional killing. In this case the dog would be a regular offer. From afterlife, he has to be an »eternal« guard for the owner's house. – (Photo U. Schmölcke).

Fig. 7 In the 2nd half of the 1st millennium AD it was popular in Northern Europe to be buried together with pets and tamed raptors. 7th-centuriy grave from Rickeby (Uppland/S). – (For details see Vretemark 2013; drawing U. Malmsteen).

fulness of the dogs but their symbolic role, which is reflected in such burials (Makiewicz 2003).

About AD 500 this tradition ended and dog skeletons disappeared from settlement contexts. Instead of this, dogs became now part of prestigious grave gifts (Prummel 1992). This new practice can be observed in Central Europe as well as in Sweden and demonstrates a strong connection between single dogs and the social prestige of their owner (**fig. 7**). These dogs were larger than the population's average, and this is especially true for Scandinavia, where often large dogs of 60-70 cm high have been buried. Potentially these animals were strong hunting dogs, because other hunting equipment is commonly also part of the grave gift (Makiewicz 2003; Schmölcke 2013). Records of huge dogs as well as written sources show that in those days a sporadic but purposeful introgression of wolves cannot be excluded (Benecke 1994, 175). Even if the large buried dogs were connected with and certainly an element of aristocratic horseback hunting, they also occur frequently in ordinary graves (Prummel 1992). Moreover, graves with a dog combined with one or more horses or a second dog are also known (Makiewicz 2003).

It can be summarised that dogs played definitely an important role for Germanics as a partner in the afterlife (Prummel 1992; Makiewicz 2003; Gräslund 2004; Scheibner 2012). Written sources, which exist in Central and Northern Europe for this period in addition to archaeological sources, reflect further facets of human's emotional life and allow us to understand at least in parts the background of the dog burial rite: they clearly show a strong connection between the dog and the death. In this connection the mythical dog was recognised as a medium between the living and the dead (Gräslund 2004). This belief was long-lasting in some parts of Europe and still known even in modern times (Lurker 1983).

It should not go unmentioned, however, that besides the spiritual significance most dogs were animals of everyday life, and this include a large variety of functions and could also comprise the use of dogs as a meat-producing animal and of the dog's pelt (Ewersen 2010).

In comparison to the Iron Age, the role of the dog changed again after Christianisation. It was still connected to the social prestige of the elite, particularly knights, but now explicitly as a valuable hunting animal and not a spiritual companion anymore (Makiewicz 2003; Schmölcke 2013). Naturally, such a high appreciation (**fig. 8**) was again aimed to a minority of dogs only (Grieve 2012). An exemplary study on dog remains from Anglo-Saxonian sites demonstrates that the large majority of dogs had hard lives (Crabtree 2015). Serving as guard and herders and involved in hunting and fighting most of the animals died relatively young and with numerous injuries and pathologies.

In the medieval Christian cultures it was not allowed anymore to bury animals in hallowed grounds, and consequently no dog skeletons can be found at cemeteries anymore. Only very few exceptional cases are known so far. Such an interesting feature is a grave excavated outside a cemetery of the 11th century at Visegrád-Vákert in Hungary (Kom. Pest/H; Vörös 1990). Here, a single female was buried in distance to all other people, and she was surrounded by six dogs. This special feature resulted in the interpretation of a »grave of a witch«. Even if we have no regular dog graves, there is a quite frequent appearance of dogs in early medieval Christian iconography and written sources, demonstrating a complex symbolic role of this

Fig. 8 In medieval times, valuable hunting dogs of the nobility were medicated and nursed. From: Gaston Phoebus. Livre de la Chasse (1387-1389). Paris, Bibliothèque National de France, Département des manuscrits, ms. Fr. 616, fol. 40v. – (agk-images, AKG357707).

species in the sacral space (Bartosiewicz 2012) as well as a rhetorical image for religious polemics (Resnick 2013).

Such a bond and level of communion observed globally between humans and dogs is unique in human-animal relationships. However, it is also true that the roles of dogs always range »from the venerated to the impure, from the tolerated to the loved« (Reece 2005, 63) – similar to today, in particular if taken in mind the situation of the feral street dogs in developing countries (Reece 2005). Always only a minority of dogs is and was handled like a companion, but large parts of the dog population lived in prehistoric and historic times on the bottom fringe of the human society and still do it today (Serpell 1995).

Returning to the relationship between the wild ancestors of dogs, wolves, and humans, we can carry forward the data and conclusions from Mesolithic and Neolithic times generally. Even if the number of sites with records of wolves and the frequency of their bones increase slightly after the change of human lifestyle from hunting-fishing-gathering to farming (Aaris-Sørensen 1977), bones of wolves remain rare in archaeological assemblages until the Early Middle Ages. Since this a supra-regional phenomenon to observe for instance in Britain, Central Europe and Scandinavia – regions, where wolves definitely occur until modern times –, some years ago a well-known archaeozoological publication was titled »Where are the Wolves?« (Pluskowski 2006a). Pluskowski's comparative study comes to the result that the limited presence of wolf remains can be explained with both infrequent hunting and the limited value of wolve's fur in relation to other fur-bearing animals. There is some evidence from Scandinavia that during the 1st millennium AD the Vikings hunted wolves for fur, but in comparison to species such as beaver, otter or small martens the wolf was of minor importance (Wigh 2001). Contemporaneous depictions show that the wolf's pelt was probably used mainly for mumming in magic and martial contexts (**fig. 9**; see Price 2016). This kind of symbolism is associated with early medieval Scandinavian poetry, which repeatedly praises great warriors as »feeders of the wolves«. Such title reflects the behaviour of wolves to appear on battlefields after the end of the fight

Fig. 9 Even if wolf pelts are not of exciting quality they have been used for coats until modern times. Picture from Umeå, Sweden, 1897. – (Photo S. Klingspor, Nordiska museet, Två män i vadlånga pälsar, Umeå, 1897 NMA.0041364).

to feast on dead soldiers (Hiltmann 2013). Based on this observation in Norse mythology wolves are mainly attributed as »producers of disorder«, but in general the picture was complex with both negative and positive connotations (Hiltmann 2013 with many details). With Fenrir, an artful opponent of the gods, and Geri and Freki, Odin's escort, this contrariness is reflected in very different roles in the Norse mythology (Simek 2006).

After Christianisation the reputation of wolves varied across Europe, but in particular in Central Europe the wolf became an allegory of bad qualities, and in medieval Christian literature wolves were commonly used as a metaphor for vice (Pluskowski 2006b). This perspective encouraged the people's motivation to kill wolves and ended not until the species was completely exterminated in this area. The reduction and ultimately extirpation of this species was official aim of the authorities in many parts of Central Europe. Often justified by the religious connotation it can be shown that at least in some regions wolf hunting was de facto driven by a completely different motif namely the wolf's disruption to the hunting culture of the nobleman (Pluskowski 2006b).

When the population density of wolves increased significantly during and after the Thirty Years' War (1618-1648), wolf hunting was intensified and officially labelled as a public duty (**fig. 10**). The official order to kill every wolf sighted was not directed because they were regarded to be dangerous for people, but because of the strong damage they made in dispatching domestic animals, in particular horses, which were of high importance as draught animals in that days (e. g. Jessen 1958; Sommer 1999). Motivated by official bounties to shoot animals and due to the purposeful destruction of nurseries of the wolves it took only 100 years to decimate the wolf population massively and only 50 more years to extirpate the species in Central Europe. Since the middle of the 19th century – now in a land without wolves – a romanticized rehabilitation occurred step by step with wolves as symbols for fidelity being the »kings of the forests« (Zimen 1978).

DOMESTICATION AND SOCIAL STATUS OF CATS

Not only today, but also in the past and at all continents humans built close relationships with felids: about 40 % of the felid species have been tamed over the millennia and in most cases the people encouraged commensal species, which had been attracted either by rodent pests or scavenging opportunities (Faure/Kitchener 2009). However, only one species was fully domesticated, probably in ancient Egypt: *Felis silvestris*, the wildcat, a highly adaptable species living in different varieties in most parts of Africa and in Eurasia from Europe to China and India (Driscoll et al. 2007). The domesticated variety, *Felis silvestris lybica*,

Fig. 10 Pit for hunting wolves build in the 17th century. Near Hohenwart (Lkr. Pfaffenhofen an der Ilm/D). – (Photo Georg Waßmuth, Wolfsgrube im Wald bei Hohenwart, CC BY-SA 3.0).

occurs in Africa and southwestern Asia. Thereby the domestic cat is in contrast to the dog a foreign faunal element in Central Europe, imported by humans.

Although present in all parts of the area since the Late Glacial, wildcats and the other European felid species, lynxes, rank among the rarest species recordable in archaeological animal bone assemblages. This applies not only for Mesolithic or Neolithic times, but also for younger periods. Obviously, both species were not hunted frequently and regularly, even if their dense winter skin was surely always a valuable resource. Due to their covert and cautious way of life and their small population density, hunting or trapping wildcats potentially had more to do with the demonstration of one's ability and with hunter's prestige (Schmitzberger 2009). Even at special purpose sites focusing on the exploitation or trading of fur bearing mammals, remains of wildcats are missing or very rare (for discussion and about the few contrary examples see Trolle-Lassen 1987; Wigh 2001, 121 ff.; Enghoff 2011, 293-298; Schmölcke/Nikulina 2015). However, convincing evidence for the human exploitation of wildcat during the Central European Stone Age only comes from the Danish Ertebølle (Nordjylland/DK), the final part of the northern Mesolithic. Particularly the site Hjerk Nor (Hjortholm/DK) yielded not only high quantities of bones from several fur bearing species, but also a large amount of remains from wildcat. A microscopic analyses of presence, number and placing of cutmarks on these bones shows that at this site besides fur also the meat of the animals was utilized regularly (Strid 2000). Traces of dismembering and filleting are characteristic and they occur in this assemblage at so many bones that eating of the wildcats is beyond doubt. However, since these special traces for the use of meat and marrow were not very desperate, humans did not use the meat intensively. Why the Stone Age people of Hjerk Nor were more interested in wildcats than people from other contemporaneous sites is difficult to explain. Hjerk Nor, however, does not change the general pattern of a subordinate relevance of the wildcat as a Stone Age hunting object.

The first evidence for a tamed cat comes from the location Khirokitia at Cyprus. 9300 years ago, a cat was buried there next to a man, on an island, where *Felis silvestris* has never occurred since the end of the Ice Age (Vigne et al. 2004). In particular the ritualistic position of the cat in the context of a human grave

Fig. 11 In Ancient Egyptian's Late Period (664-332 BC) an unknown artist created this about 40cm high statue of a cat made of bronze with gold ornaments, the so called Gayer-Anderson Cat. It probably represents the goddess Bastet. She is also known as a woman with the head of a cat (Clutton-Brock 2000). – (Photo Einsamer Schütze, British Museum Egypt 101-black, CC BY-SA 3.0).

suggests a cultural importance of cats and a kind of spiritual connection of the owner to this special animal (Vigne et al. 2004).

It is plausible that cats participated actively in their own domestication, and that wildcats subsequently tolerated direct human presence from the time of Khirokitia onward (Lyons/Kurushima 2012). Potentially, for millennia commensal or tamed cats lived in association with humans, and probably this domestication process occurred in several regions independently (Bar-Oz/Weissbrod/Tsahar 2014; Hu et al. 2014; Montague et al. 2014).

In Egypt between 4000 and 3900 years ago during the 12th Dynasty people lived together with cats, and regular breeding seems to have been started some millennia later (Linseele/Van Neer/Hendrickx 2007; Kurushima et al. 2012; Van Neer et al. 2014). Cats played an important role in Egyptian spiritual life and in grave-like contexts a great number of mummified cats is known, so that ancient Egyptians could be regarded as »cat people« with cats as one of the most sacred animals (Morey 2006). The buried cats were larger than their wild ancestors, so they must have had very good living conditions. The high status and reputation of cats is impressively reflected in Bastet, the goddess of joy and dance, the protectress of pregnant women (**fig. 11**; Malek 1997; Schuller 2007). The Bastet cult likely began during the 22nd Dynasty (about 945-715 BC), and from the writings of the Greek historian Herodotus we are well informed about all details of the ceremonies during the zenith of the cult in the 5th century BC (Serpell 2014).

However, this level of regard was a local Egyptian matter, and in ancient Greece and Rome cat did not play a prominent role. Tame wildcats or domestic cats were known in Greece since about 1400 BC and in Italy from 500 BC onwards (for details see Faure/Kitchener 2009). However, the spread of cats throughout the Mediterranean was slow, because ancient Egyptians forbade the export of cats (Faure/Kitchener 2009). It is remarkable that, following the iconography, in classical Greece cats were already used as pets (Malek 1997). Finally, attendant to the Romans, domestic cats also reached Central Europe during the first centuries AD (Benecke 1994, 145ff.), but during the following centuries their bones remain always rare in archaeological assemblages from this area. A typical example for this time is Reric, today Groß Strömkendorf (Lkr. Nordwestmecklenburg/D), near Wismar in northern Germany, settled between AD 730 and 800, where from more than 25000 bones from domesticated animal species only 19 cat remains have been found (0.08%; Schmölcke 2004, 70ff.). Only one to two centuries later, at the end of the 1st millennium AD, in one of the most important settlements of the Vikings, Hedeby (the beneficiary of Reric's destruction; Schmölcke 2004), bones of cats accounted already 0.4% of the hundreds of thousands remains of domestic animals (Johansson/Hüster 1987; Heinrich 1996). It was after the year AD 1000 that cat remains in archaeological

Fig. 12 About 1618 in England the so-called Witches of Belvoir were three women, a mother and her two daughters, accused of witchcraft. Their cats were important attributes to them. – (The Wonderful Discoverie of the Witchcrafts of Margaret and Phillip Flower.... Together with the seuerall Examinations and Confessions of Anne Baker, Ioan Willimot, and Ellen Greene, Witches in Leicestershire [London 1619] 1).

assemblages from Central European settlements become rapidly and generally more frequent, potentially because large mice and rat populations in the growing towns provided rich dietary resources. An example for this development offers the border region between Germany and Denmark. In the medieval town of Schleswig (Kr. Schleswig-Flensburg/D), the successor of the directly neighboured Hedeby, cats have an account of 2.4 % of bones from domesticated animals – i. e. more than six times more than in Hedeby and 30 times more than in Reric some hundred years before (Heinrich 1996; Schmölcke 2004, fig. 47). In contrast to its ancestor Hedeby, in medieval Schleswig lived much more cats than dogs (Spahn 1986, 45). Surely, most of these cats were kept for mousing, but at least in some medieval towns this increase also coincides with increasing skinning records (Serjeantson 1989).

Based on multidisciplinary data, K. Poole has investigated the cat-human relationship between the 5th and the 11th century in detail (Poole 2014). He mentions different written sources showing that the people perceived mice-hunting as the function, duty or »job« of a cat and that successful and effective mousers could reach the same value as a woolly sheep with a lamb. The fact that at least some cats had a name shows that they were recognised as individuals with own specific character. However, at the same time cats were also seen as a source of fur, although it is not clear if they were killed purposeful or if they were found dead before skinned. Poole points out that in assemblages of animal remains from early medieval times cat remains with cutmarks are much more frequent on bones from young animals, indicating a preference for the fur of young cats. The reason for this might be the fact that young cat fur is softer. In Central Europe, for example at Hedeby, many cats also died at the age of nine to twelve months (Johansson/Hüster 1987). Thus, given a birth in spring, they died during their first winter. It could be suspected that these cats have been killed by humans to use their dense and soft winter pelts, but at Hedeby only single cat remains derive cutmarks. However, even if in this special case the age distribution probably reflects a high mortality of young cats due to underfeeding or diseases rather than direct human killing, this result is congruent with K. Poole's statement that in early medieval times many humans did not feed the cats regularly, to provide an incentive to mice hunting.

Latest about AD 1300 cats became to be regarded as »creatures of the devil« in large parts of Central Europe (Clutton-Brock 2000; Thomas 2005; von Heusinger 2007). Already for Hildegard of Bingen, cats were

toxic organisms (in *Physica* 7,26; written about AD 1150), and in the work *Opus de natura rerum* written by the medieval theologian Thomas of Cantimpré between 1225 and 1241 the cat epitomises several sinful features (Meier 2008, 62 ff.). Especially black cats were believed to be incarnations of evil, origin of plagues and accidents and companion of witches (**fig. 12**), and as a result cats were killed on a large scale. As a consequence, cats became so rare in Central European towns and villages that the populations of mice and rats increased rapidly in the course of the 14th century (Oeser 2004, 88-105; Hengerer 2009). Due to several facets of their behaviour – their demonstrative independency, their intraspecific intolerance, their nocturnal screeching – for many medieval people cats were not social beings such as dogs, but noticed as latent dangerous and antisocial. Cats »became a metaphor for female sexual depravity and social unruliness, and were persecuted and despised« (Serpell 2014, 100). Examinations of paintings and written sources show that their reputation only slowly changed (Hengerer 2007; 2009). However, there were also positive aspects in the relation between cats and humans: Since these animals were hold to be especially fertile and reproductive, they played a special role in matters of love. In parts of Europe a cat was a common wedding present, expressing the wish of many children for the newly married pair (Hengerer 2007).
It was only during the 19th century when cats started to become fashionable middle-class pets. Simultaneously the human attitude towards cats changed towards a predominantly positive view (Serpell 2014). In that days and today in large parts of Europe and North America cats have replaced dogs as the most popular animal species (Turner/Bateson 2014, tab. 1.1; Turner/Waiblinger/Fehlbaum 2014). However, even in Central Europe there are still people who connect to a sudden appearance of a black cat with bad luck, and others see them as furtive and untrustworthy (Serpell 2014). In contrast to dogs, even after millennia of domestication the cat behaviour still exhibits the biology of the wild ancestor.

CONCLUSIONS

In former human-carnivores relationships, the animals often had first and foremost a symbolic and non-rational significance. This factor can be demonstrated for the wild species bear and wolf, but also for dog and domestic cat. People connected several spiritual aspects with these animals, e. g. their relevance as co-workers and companions in life and afterlife, their role as agents between the here and the beyond or generally their position as mediators between humans and nature. In comparison to these aspects, connected to people's belief and their world of faith, more practical issues such as hunting assistance or mousing seem to have played a minor role. Surely, this was also important, but often subordinated and beneficial.
Such symbolic characters built a continuous background to observe consistently in different cultures, regions and times. Against this background a conspicuous change in the reputation of carnivores took place in the course of the last 10000 years of human history. These changes always mirror contemporaneous shifts in human's way of life and in religious and spiritual beliefs. They can be detected in prehistoric and historic times as well as in recent days.
People associated and personalised elementary forces of nature, such as death and power over the lives of others particularly with the large wild carnivores bear and wolf in prehistory and early historic times. Later, in the Christian Middle Age, the belief of the presence of devil, who can appear as an animal, and of witches, accompanied by diabolic creatures, strongly influenced the reputation of carnivores, both wild and domesticated. Accordingly, the human relationship to animals, in particular to carnivores, was always part of the regnant world view or world concept in the past. This still holds true today, were the reputation of carnivores has changed again in large parts of Europe. Due to environmental problems, the extinction of

species and decreasing populations of many animals and plants, ideas to protect environment and species are relevant to society. In this modern perspective bears and wolves are decisive indicator species for intact ecosystems. Along with the spread of the idea to protect nature, this aspect becomes manifest in strong international statutes for conservation of species. Thereby, at least in many part of Europe in the changeful history of human-carnivores-interactions a new chapter is opened.

REFERENCES

Aaris-Sørensen 1977: K. Aaris-Sørensen, The subfossil wolf *Canis lupus* in Denmark. Videnskabelige Meddelelser fra Dansk Naturhistorisk Forening 140, 1977, 129-146.

Bar-Oz/Weissbrod/Tsahar 2014: G. Bar-Oz / L. Weissbrod / E. Tsahar, Cats in recent Chinese study on cat domestication are commensal, not domesticated. Proceedings of the National Academy of Sciences of the United States of America 111, 2014, E876. DOI: 10.1073/pnas.1324177111.

Bartosiewicz 2009: L. Bartosiewicz, A Lion's Share of Attention: Archaeozoology and the Historical Record. Acta Archaeologica Academiae Scientiarum Hungaricae 60, 2009, 275-289.

2012: L. Bartosiewicz, »Stone Dead«: Dogs in a Medieval Sacral Space. In: A. Pluskowski (ed.), The Ritual Killing and Burial of Animals: European Perspectives (Oxford 2012) 220-229.

Benecke 1987: N. Benecke, Studies on early dog remains from Northern Europe. Journal of Archaeological Science 14, 1987, 31-49.

1994: N. Benecke, Archäozoologische Studien zur Entwicklung der Haustierhaltung in Mitteleuropa und Südskandinavien von den Anfängen bis zum ausgehenden Mittelalter. Schriften zur Ur- und Frühgeschichte 46 (Berlin 1994).

2000: N. Benecke, Die jungpleistozäne und holozäne Tierwelt Mecklenburg-Vorpommerns. Beiträge zur Ur- und Frühgeschichte Mitteleuropas 23 (Weissbach 2000).

Bieder 2005: R. E. Bieder, Bear (London 2005).

Bökönyi 1974: S. Bökönyi, History of Domestic Mammals in Central and Eastern Europe (Budapest 1974).

Boyle 2006: K. Boyle, Neolithic Wild Game Animals in Western Europe: the Question of Hunting. In: D. Serjeantson / D. Field (eds), Animals in the Neolithic of Britain and Europe (Oxford 2006) 10-23.

Brantingham 1998: P. J. Brantingham, Hominid-carnivore coevolution and invasion of the predatory guild. Journal of Anthropological Archaeology 17, 1998, 327-353.

Brunner 2010: B. Brunner, Bär und Mensch. Die Geschichte einer Beziehung (Darmstadt 2010).

Bürglin 2015: R. Bürglin, Sie sind wieder da. Bär, Luchs und Wolf erleben (Stuttgart 2015).

Butzeck/Stubbe/Piechocki 1988: S. Butzeck / M. Stubbe / R. Piechocki, Beiträge zur Geschichte der Säugetierfauna der DDR. 1: Der Braunbär *Ursus arctos* Linné 1758. Hercynia 25, 1988, 27-59.

Černych 1999: P. Ja. Černych, Историко-этимологический словарь русского языка: в 2-х т. Т. 1 (Moskva 1999).

von Chamier 2017: R. von Chamier, Hunde essen, Hunde lieben. Die Tabugeschichte des Hundeverzehrs und das erstaunliche Kapitel deutscher Hundeliebe (Baden-Baden 2017).

Charles 1997: R. Charles, The Exploitation of Carnivores and Other Fur-bearing Mammals during the North-western European Late and Upper Paleolithic and Mesolithic. Oxford Journal of Archaeology 16, 1997, 253-277.

Clutton-Brock 1995: J. Clutton-Brock, Origins of the dog: domestication and early history. In: J. Serpell (ed.), The domestic dog: its evolution, behaviour, and interactions with people (London 1995) 7-20.

1999: J. Clutton-Brock, A natural history of domesticated mammals (Cambridge 1999).

2000: J. Clutton-Brock, The British Museum Book of Cat (London 2000).

Crabtree 2015: P. J. Crabtree, A Note on the Role of Dogs in Anglo-Saxon Society: Evidence from East Anglia. International Journal of Osteoarchaeology 25, 2015, 976-980.

Crockford 2000: S. J. Crockford (ed.), Dogs Through Time: An Archaeological Perspective. Proceedings of the 1st ICAZ Symposium on the History of the Domestic Dog; Eighth Congress of the International Council for Archaeozoology (ICAZ98), August 23-29, 1998, Victoria, BC, Canada. British Archaeological Reports: International Series 889 (Oxford 2000).

Deinet et al. 2013: S. Deinet / C. Ieronymidou / L. McRae / I. J. Burfield / R. P. Foppen / B. Collen / M. Böhm, Wildlife comeback in Europe: the recovery of selected mammal and bird species (London 2013).

Diedrich 2014: C. G. Diedrich, Palaeopopulations of Late Pleistocene Top Predators in Europe: Ice Age Spotted Hyenas and Steppe Lions in Battle and Competition about Prey. Paleontology Journal 2014, 1-34. DOI: 10.1155/2014/106203.

Driscoll et al. 2007: C. A. Driscoll / M. Menotti-Raymond / A. L. Roca / K. Hupe / W. E. Johnson / E. Geffen / E. H. Harley / M. Delibes / D. Pontier / A. C. Kitchener / N. Yamaguchi / S. J. O'Brien / D. W. Macdonald, The Near Eastern origin of cat domestication. Science 317, 2007, 519-523.

Duerr 2010: J. Duerr, Von Tierhütern und Tiertötern. Mythos und Ethik der Jagd im kulturhistorischen Vergleich. Universitätsforschungen zur prähistorischen Archäologie 189 (Bonn 2010).

Enghoff 2011: I. B. Enghoff, Regionality and biotope exploitation in Danish Ertebolle and adjoining periods. Scientia Danica: Series B, Biologica 1 (Copenhagen 2011).

Ewersen 2010: J. Ewersen, Hundehaltung auf der kaiserzeitlichen Wurt Feddersen Wierde – ein Rekonstruktionsversuch / Dog husbandry at the Roman Iron Age dwelling mound Feddersen Wierde – an attempt at reconstruction. In: Herrenhöfe und die Hierarchie der Macht im Raum südlich und östlich der Nordsee von der vorrömischen Eisenzeit bis zum frühen Mittelalter und zur Wikingerzeit; 11.-13. Oktober 2007, Burg Bederkesa in Bad Bederkesa. Siedlungs- und Küstenforschung im südlichen Nordseegebiet 33 (Rahden/Westf. 2010) 53-75.

Ewersen/Ramminger 2013: J. Ewersen / B. Ramminger, Zur Haltung und Nutzung von Haushunden auf neolithischen Fundplätzen in Mittel- und Süddeutschland sowie der Schweiz. Germania 91, 2013, 1-37.

Ewersen et al. 2018: J. Ewersen / S. Ziegler / B. Ramminger / U. Schmölcke, Stable isotopic ratios from Mesolithic and Neolithic canids as an indicator of human economic and ritual activity. Journal of Archaeological Science: Reports 17, 2018, 346-357.

Fascione/Delach/Smith 2004: N. Fascione / A. Delach / M. Smith, People and predators: from conflict to coexistence (Washington 2004).

Fahlander 2008: F. Fahlander, A piece of the Mesolithic. Horizontal stratigraphy and bodily manipulations at Skateholm. In: F. Fahlander / T. Oestigaard, The Materiality of Death: Bodies, Burials, Beliefs. British Archaeological Reports: International Series 1768 (London 2008) 29-46.

Faure/Kitchener 2009: E. Faure / A. C. Kitchener, An Archaeological and Historical Review of the Relationships between Felids and People. Anthrozoos 22, 2009, 221-238.

Feeser et al. 2012: I. Feeser / W. Dörfler / F. R. Averdiek / J. Wiethold, New insight into regional and local land-use and vegetation patterns in eastern Schleswig-Holstein during the Neolithic. In: M. Hinz / J. Müller (eds), Siedlung, Grabenwerk, Großsteingrab. Studien zu Gesellschaft, Wirtschaft und Umwelt der Trichterbechergruppen im nördlichen Mitteleuropa. Frühe Monumentalität und soziale Differenzierung 2 (Bonn 2012) 159-190.

Geppert 1990: P. Geppert, Hundeschlachtungen in Deutschland im 19. und 20. Jahrhundert: unter besonderer Berücksichtigung der Verhältnisse in München [unpubl. PhD thesis Univ. Munich 1990].

Germonpré et al. 2009: M. Germonpré / M. V. Sablin / R. E. Stevens / R. E. M. Hedges / M. Hofreiter / M. Stiller / V. R. Després, Fossil dogs and wolves from Palaeolithic sites in Belgium, the Ukraine and Russia: osteometry, ancient DNA and stable isotopes. Journal of Archaeological Science 36, 2009, 473-490.

Gramsch 2014: B. Gramsch, Die Schmuckfunde von Friesack, Fundplatz 4, Lkr. Havelland. Veröffentlichungen zur brandenburgischen Landesarchäologie 46, 2012 (2014), 7-26.

Gräslund 2004: A. S. Gräslund, Dogs in graves – a question of symbolism? In: B. Santillo Frizell (ed.), Man and animal in antiquity. Proceedings of the Conference at the Swedish Institute in Rome, September 9-12, 2002. Swedish Institute: Projects and Seminars 1 (Rom 2004) 171-180.

Grieve 2012: A. Grieve, The human-dog relationship in early medieval England and Ireland (c. AD 400-1250) [unpubl. PhD thesis Univ. Southampton 2012].

Grimm 2013: O. Grimm, Bear-skins in northern European burials and some remarks on other bear-related furnishings in the north and middle of Europe in the 1st millennium AD. In: O. Grimm / U. Schmölcke (eds), Hunting in Northern Europe until 1500 AD. Old traditions and regional developments, continental sources and continental influences. Schriften des Archäologischen Landesmuseums: Ergänzungsreihe 7 (Neumünster 2013) 277-296.

Grimm/Schmölcke 2013: O. Grimm / U. Schmölcke, Results and future perspectives in relation to an overall concept of hunting-related research. In: O. Grimm / U. Schmölcke (eds), Hunting in Northern Europe until 1500 AD. Old traditions and regional developments, continental sources and continental influences. Schriften des Archäologischen Landesmuseums: Ergänzungsreihe 7 (Neumünster 2013) 605-631.

Grimm et al. 2014: O. Grimm / H. Lübke / H. Paulsen / U. Schmölcke / M.-J. Weber, Jagdgeschichte an einigen Beispielen aus Norddeutschland – Von der Rentierjagd zur Beizjagd. Archäologische Nachrichten aus Schleswig-Holstein 20, 2014, 22-29.

Hartz/Schmölcke 2013: S. Hartz / U. Schmölcke, From the Mesolithic to the Neolithic – Stone Age hunting strategies in the southwestern Baltic Sea area. In: O. Grimm / U. Schmölcke (eds), Hunting in Northern Europe until 1500 AD. Old traditions and regional developments, continental sources and continental influences. Schriften des Archäologischen Landesmuseums: Ergänzungsreihe 7 (Neumünster 2013) 21-40.

Hartz et al. 2014: S. Hartz / H. Jöns / H. Lübke / U. Schmölcke / C. von Carnap-Bornheim / D. Heinrich / S. Klooss / F. Lüth / S. Wolters, Prehistoric Settlements in the South-western Baltic Sea area and Development of the Regional Stone Age Economy. Bericht der Römisch-Germanischen Kommission 92, 2011 (2014), 77-210.

Hauck 1963: K. Hauck, Tiergärten im Pfalzbereich. In: A. Gauert (ed.), Deutsche Königspfalzen. Beiträge zu ihrer historischen und archäologischen Erforschung 1. Veröffentlichungen des Max-Planck-Instituts für Geschichte 11, 1 (Göttingen 1963) 30-74.

Heinrich 1996: D. Heinrich, Die Faunenreste von Schleswig und anderer mittelalterlicher Siedlungsplätze Schleswig-Holsteins. Eine vergleichende Betrachtung. Offa 53, 1996, 327-370.

Hemmer 1990: H. Hemmer, Domestication: the decline of environmental appreciation (Cambridge 1990).

Hengerer 2007: M. Hengerer, Die Katze in der Frühen Neuzeit: Stationen auf dem Weg zur Seelenverwandten des Menschen. In: C. Wischenmann (ed.), Von Katzen und Menschen: Sozialgeschichte auf leisen Sohlen (Konstanz 2007) 53-88.

2009: M. Hengerer, Stadt, Land, Katze. Zur Geschichte der Katze in der Frühneuzeit. Informationen zur modernen Stadtgeschichte 2, 2009, 13-25.

Herre/Röhrs 1990: W. Herre / M. Röhrs, Haustiere – zoologisch gesehen (Stuttgart 21990).

von Heusinger 2007: S. von Heusinger, Die Katze im Kochtopf. Zu Ernährung und Kultur des Mittelalters. In: C. Wischenmann (ed.), Von Katzen und Menschen: Sozialgeschichte auf leisen Sohlen (Konstanz 2007) 33-52.

Hiltmann 2013: H. Hiltmann, »Guter Wolf – böser Wolf«: Die Ambivalenz norröner Wolfsbilder und ihrer persönlichkeitsstiftenden Funktion im europäischen Vergleich. In: M. Ulrich / D. D. Rentiis (eds), Animalia in fabula. Interdisziplinäre Gedanken über das Tier in der Sprache, Literatur und Kultur. Schriften aus der Fakultät Geistes- und Kulturwissenschaften der Otto-Friedrich-Universität Bamberg 14 (Bamberg 2013) 143-174.

Horard-Herbin/Tresset/Vigne 2014: M. P. Horard-Herbin / A. Tresset / J. D. Vigne, Domestication and uses of the dog in western

Europe from the Paleolithic to the Iron Age. Animal Frontiers 4, 2014, 23-31. DOI: 10.2527/af.2014-0018.

Hu et al. 2014: Y. Hu / S. Hu / W. Wang / X. Wu / F. B. Marshall / X. Chen / L. Hou / C. Wang, Earliest evidence for commensal processes of cat domestication. Proceedings of the National Academy of Sciences of the United States of America 111, 2014, 116-120. DOI: 10.1073/pnas.1311439110.

Jessen 1958: H. Jessen, Jagdgeschichte Schleswig-Holsteins (Rendsburg 1958).

Johansson/Hüster 1987: F. Johansson / H. Hüster, Untersuchungen an Skelettresten von Katzen aus Haithabu (Ausgrabung 1966-1969). Berichte über die Ausgrabungen in Haithabu 24 (Neumünster 1987).

Kalb 2007: R. Kalb, Bär, Luchs, Wolf – verfolgt, ausgerottet, zurückgekehrt (Graz, Stuttgart 2007).

Kim 2007: R. E. Kim, Dog meat in Korea: a socio-legal challenge. Animal Law Review 14, 2007, 201-236.

Kindler 2012: L. Kindler, Die Rolle von Raubtieren bei der Einnischung und Subsistenz jungpleistozäner Neandertaler: Archäozoologie und Taphonomie der mittelpaläolithischen Fauna aus der Balver Höhle (Westfalen). Monographien des RGZM 99 (Mainz 2012).

Kluge/von Seebold 2012: F. Kluge / E. von Seebold, Etymologisches Wörterbuch der deutschen Sprache (Berlin [25]2012).

Kruuk 2002: H. Kruuk, Hunter and Hunted. The Relationship between Carnivores and People (Cambridge 2002).

Kurushima et al. 2012: J. D. Kurushima / M. J. Lipinski / B. Gandolfi / L. Froenicke / J. C. Grahn / R. A. Grahn / L. A. Lyons, Variation of cats under domestication: genetic assignment of domestic cats to breeds and worldwide random-bred populations. Animal Genetics 44, 2012, 311-324.

Larson et al. 2012: G. Larson / E. K. Karlsson / A. Perri / M. T. Webster / S. Y. Ho / J. Peters / P. W. Stahl / P. J. Piper / F. Lingaas / M. Fredholm / K. E. Comstock / J. F. Modiano / C. Schelling / A. I. Agoulnik / P. A. Leegwater / K. Dobney / J. D. Vigne / C. Vilà / L. Andersson / K. Lindblad-Toh, Rethinking dog domestication by integrating genetics, archeology, and biogeography. Proceedings of the National Academy of Sciences of the United States of America 109, 2012, 8878-8883. DOI: 10.1073/pnas.1203005109.

Larsson 1990. L. Larsson, Dogs in fraction – symbols in action. In: P. M. Vermeersch / P. Van Peer (eds), Contributions to the Mesolithic in Europe. Papers presented at the fourth international symposium »Mesolithic in Europe«. Studia Praehistorica Belgica 5 (Leuven 1990) 153-160.

Lehmkuhl 2010: U. Lehmkuhl, Die Tierbeigaben aus den Gräbern von Ostorf (Mecklenburg) unter besonderer Berücksichtigung der durchbohrten Tierzähne. Mit einem Beitrag zu den Tierbeigaben aus den Gräbern von Tangermünde bei Stendal. Bericht der Römisch-Germanischen Kommission 88, 2007 (2010), 385-414.

Lindner 1940: K. Lindner, Geschichte des deutschen Weidwerks. 2: Die Jagd im frühen Mittelalter (Berlin 1940).

Linseele/Van Neer/Hendrickx 2007: V. Linseele / W. Van Neer / S. Hendrickx, Evidence for early cat taming in Egypt. Journal of Archaeological Science 34, 2007, 2081-2090.

Lüning 2000: J. Lüning, Steinzeitliche Bauern in Deutschland. Die Landwirtschaft im Neolithikum. Universitätsforschungen zur prähistorischen Archäologie 58 (Bonn 2000).

Lurker 1983: M. Lurker, Der Hund als Symboltier für den Übergang vom Diesseits in das Jenseits. Zeitschrift für Religions- und Geistesgeschichte 35, 1983, 132-144.

Lyons/Kurushima 2012: L. A. Lyons / J. D. Kurushima, A Short Natural History of the Cat and Its Relationship with Humans. In: S. Little (ed.), The Cat: Clinical Medicine and Management (Philadelphia PA 2012)1254-1262.

Makiewicz 2003: T. Makiewicz, The changing role and significance of dogs for Germanic people from the beginning of the Roman Period till the Middle Ages. Archeozoologia 21, 2003, 88-102.

Malek 1997: J. Malek, The cat in ancient Egypt (Philadelphia PA 1997).

Manwell/Baker 1984: C. Manwell / C. M. Baker, Domestication of the dog: hunter, food, bed-warmer, or emotional object? Zeitschrift für Tierzüchtung und Züchtungsbiologie 101, 1984, 241-256.

Meier 2008: F. Meier, Mensch und Tier im Mittelalter (Ostfildern 2008).

Molsdorf 1926: W. Molsdorf, Christliche Symbolik der mittelalterlichen Kunst. Hiersemanns Handbücher 10 (Leipzig [2]1926).

Montague et al. 2014: M. J. Montague / G. Li / B. Gandolfi / R. Khan / B. L. Aken / S. M. Searle / P. Minx / L. W. Hillier / D. C. Koboldt / B. W. Davis / C. A. Driscoll / Ch. S. Barr / K. Blackistone / J. Quilez / B. Lorente-Galdos / T. Marques-Bonet / C. Alkan / G. W. C. Thomas / M. W. Hahn / M. Menotti-Raymond / S. J. O'Brien / R. K. Wilson / L. A. Lyons / W. J. Murphy / W. C. Warren, Comparative analysis of the domestic cat genome reveals genetic signatures underlying feline biology and domestication. Proceedings of the National Academy of Sciences of the United States of America 111, 2014, 17230-17235. DOI: 10.1073/pnas.1410083111.

Morey 1992: D. F. Morey, Size, shape and development in the evolution of the domestic dog. Journal of Archaeological Science 19, 1992, 181-204.

2006: D. F. Morey, Burying key evidence: the social bond between dogs and people. Journal of Archaeological Science 33, 2006, 158-175.

2010. D. F. Morey, Dogs: domestication and the development of a social bond (Cambridge 2010).

2014: D. F. Morey, In search of Paleolithic dogs: a quest with mixed results. Journal of Archaeological Science 52, 2014, 300-307.

Müller 1971: H.-H. Müller, Der Braunbär – *Ursus arctos* L. – im Mittelalter. Hercynia 8, 1971, 52-57.

Nobis 1986: G. Nobis, Die Wildsäugetiere in der Umwelt des Menschen von Oberkassel bei Bonn und das Domestikationsproblem von Wölfen im Jungpaläolithikum. Bonner Jahrbücher 186, 1986, 368-376.

Oehrl 2013: S. Oehrl, Svá beitum vér björnuna á mörkinni norðr – Bear hunting and its ideological context (as a background for the interpretation of bear claws and other remains of bears in Germanic graves of the 1st millennium AD). In: O. Grimm / U. Schmölcke (eds), Hunting in Northern Europe until 1500 AD. Old traditions and regional developments, continental sources and continental influences. Schriften des Archäologischen Landesmuseums: Ergänzungsreihe 7 (Neumünster 2013) 297-332.

Oeser 2004: E. Oeser, Katze und Mensch. Die Geschichte einer Beziehung (Darmstadt 2004).

Otterstedt/Rosenberger 2012: C. Otterstedt / M. Rosenberger (eds), Gefährten – Konkurrenten – Verwandte: Die Mensch-Tier-Beziehung im wissenschaftlichen Diskurs (Göttingen 2012).

Paul 1981: M. Paul, Wolf, Fuchs und Hund bei den Germanen. Wiener Arbeiten zur germanischen Altertumskunde und Philologie 13 (Wien 1981).

Peters 1997: J. Peters, Der Hund in der Antike aus archäozoologischer Sicht. Anthropozoologica 25/26, 1997, 511-523.

Petré 1980: B. Petré, Björnfällen i begravningsritualen – statusobjekt speglande regional skinnhandel? Fornvännen 75, 1980, 5-14.

Pluskowski 2006a: A. Pluskowski, Where are the wolves? Investigating the scarcity of European grey wolf (*Canis lupus lupus*) remains in medieval archaeological contexts and its implications. International Journal of Osteoarchaeology 16, 2006, 279-295.

2006b: A. Pluskowski, Wolves and the Wilderness in the Middle Ages (Woodbridge 2006).

Pokorny 1959: J. Pokorny, Indogermanisches Etymologisches Wörterbuch (Tübingen 1959).

Poole 2014: K. Poole, The Contextual Cat: Human–Animal Relations and Social Meaning in Anglo-Saxon England. Journal of Archaeological Method and Theory 22, 2014, 857-882.

Price 2016: N. Price, The Viking Way: Religion and War in Late Iron Age Scandinavia (Oxford 2016).

Prummel 1992: W. Prummel, Early medieval dog burials among the germanic tribes. Helinium 32, 1992, 132-194.

Reece 2005: J. F. Reece, Dogs and dog control in developing countries. In: D. J. Salem / A. N. Rowan (eds), The state of the animals III (Washington 2005) 55-64.

Resnick 2013: I. M. Resnick, Good Dog/Bad Dog: Dogs in Medieval Religious Polemics. Enarratio: Publications of the Medieval Association of the Midwest 18, 2013, 70-97.

Rosell/Blasco 2009: J. Rosell / R. Blasco, Home sharing: carnivores in anthropogenic assemblages of the Middle Pleistocene. Journal of Taphonomy 7, 2009, 305-324.

Rosell et al. 2012: J. Rosell / E. Baquedano / R. B. López / E. Camarós, New insights on Hominid-Carnivore interactions during the Pleistocene. Journal of Taphonomy 10, 2012, 125-128.

Russell 2012: N. Russell, Social zooarchaeology: Humans and animals in prehistory (Cambridge 2012).

Scheibner 2012: A. Scheibner, Des Menschen bester Freund? Die Rolle des Hundes in der Vorrömischen Eisenzeit im deutsch-polnischen Raum. Mitteilungen der Berliner Gesellschaft für Anthropologie, Ethnologie und Urgeschichte 33, 2012, 1-16.

Schleidt/Shalter 2003: W. M. Schleidt / M. D. Shalter, Co-evolution of humans and canids. Evolution and Cognition 9, 2003, 57-71.

Schmitzberger 2009: M. Schmitzberger, Haus- und Jagdtiere im Neolithikum des österreichischen Donauraumes [unpubl. PhD thesis Univ. Vienna].

Schmölcke 2004: U. Schmölcke, Nutztierhaltung, Jagd und Fischerei. Zur Nahrungsmittelwirtschaft des frühgeschichtlichen Handelsplatzes von Groß Strömkendorf, Landkreis Nordwestmecklenburg (Lübstorf 2004).

2013: U. Schmölcke, The evidence of dogs for hunting from Mesolithic times up to the Viking Age from a zoological point of view – A survey. In: O. Grimm / U. Schmölcke (eds), Hunting in Northern Europe until 1500 AD. Old traditions and regional developments, continental sources and continental influences. Schriften des Archäologischen Landesmuseums: Ergänzungsreihe 7 (Neumünster 2013) 175-183.

Schmölcke/Nikulina 2015: U. Schmölcke / E. A. Nikulina, Mesolithic beaver hunting station or base camp of supra-regional Stone Age fur trade? New archaeozoological and archaeogenetic results from Dąbki 9. In: J. Kabaciński / S. Hartz / D. C. M. Raemaekers / Th. Terberger (eds), The Dąbki Site in Pomerania and the Neolithisation of the North European Lowlands (c. 5000-3000 calBC) (Rahden/Westf. 2015) 65-86.

Schmölcke/Groß/Nikulina 2017: U. Schmölcke / D. Groß / E. A. Nikulina, Bears and beavers. »The Browns« in daily life and spiritual world. In: B. V. Eriksen / A. Abegg-Wigg / R. Bleile / U. Ickerodt (eds.), Interaction without borders. Exemplary archaeological research at the beginning of the 21st century (Schleswig 2017) 901-916.

Schuller 2007: W. Schuller, Bastet, das Kätzchen. Die Katze im alten Ägypten. In: C. Wischenmann (ed.), Von Katzen und Menschen: Sozialgeschichte auf leisen Sohlen (Konstanz 2007) 13-23.

Serjeantson 1989: D. Serjeantson, Animal remains and the tanning trade. In: D. Serjeantson / T. Waldron (eds), Diet and Crafts in Towns. The Evidence of Animal Remains from the Roman to the Post-Medieval Periods (Oxford 1989) 129-146.

Serpell 1995: J. A. Serpell, From paragon to pariah: some reflections on human attitudes to dogs. In: J. A. Serpell (ed.), The Domestic Dog: Its Evolution, Behaviour & Interactions with People (Cambridge 1995) 246-256.

2014: J. A. Serpell, Domestication and history of the cat. In: D. C. Turner / P. P. G. Bateson (eds), The domestic cat: the biology of its behaviour (Cambridge 2014) 83-100.

Simek 2006: R. Simek, Lexikon der germanischen Mythologie (Stuttgart 2006).

Skoglund et al. 2015: P. Skoglund / E. Ersmark / E. Palkopoulou / L. Dalen, Ancient wolf genome reveals an early divergence of domestic dog ancestors and admixture into high-latitude breeds. Current Biology 25, 2015, 1515-1519.

Sommer 1999: R. Sommer, Der Wolf in Mecklenburg-Vorpommern: Vorkommen und Geschichte (Schwerin 1999).

Sommer/Benecke 2005: R. S. Sommer / N. Benecke, Late-Pleistocene and early Holocene history of the canid fauna of Europe (Canidae). Mammalian Biology – Zeitschrift für Säugetierkunde 70, 2005, 227-241.

2006: R. S. Sommer / N. Benecke, Late Pleistocene and Holocene development of the felid fauna (Felidae) of Europe: a review. Journal of Zoology 269, 2006, 7-19.

Spahn 1986: N. Spahn, Untersuchungen an Skelettresten von Hunden und Katzen aus dem mittelalterlichen Schleswig, Ausgrabung Schild 1971-1975. Ausgrabungen in Schleswig, Berichte und Studien 5 (Nemünster 1986).

Steffens 2005: J. Steffens, Bedeutung der Jagd in der TBK. Journal of Neolithic Archaeology 7, 2005, 1-12.

Stiner 2002: M. C. Stiner, Carnivory, coevolution, and the geographic spread of the genus Homo. Journal of Archaeological Research 10, 2002, 1-63.

Street 2002: M. Street, Ein Wiedersehen mit dem Hund von Bonn-Oberkassel. Bonner Zoologische Beiträge 50, 2002, 269-290.

Strid 2000: L. Strid, To eat or not to eat? The significance of the cut marks on the bones from wild canids, mustelids and felids from the Danish Ertebølle site Hjerk Nor [unpubl. MA thesis Univ. Southampton].

Stuart 2015: A. J. Stuart, Late Quaternary megafaunal extinctions on the continents: a short review. Geological Journal 50, 2015, 338-363.

Stuart/Lister 2011: A. J. Stuart / A. M. Lister, Extinction chronology of the cave lion *Panthera spelaea*. Quaternary Science Reviews 30, 2011, 2329-2340.

2014: A. J. Stuart / A. M. Lister, New radiocarbon evidence on the extirpation of the spotted hyaena (*Crocuta crocuta* (Erxl.)) in northern Eurasia. Quaternary Science Reviews 96, 2014, 108-116.

Thalmann et al. 2013: O. Thalmann / B. Shapiro / P. Cui / V. J. Schuenemann / S. K. Sawyer / D. L. Greenfield / M. B. Germonpré / M. V. Sablin / F. Lopez-Giraldez / X. Domingo-Roura / H. Napierala / H. P. Uerpmann / D. M. Loponte / A. A. Acosta / L. Giemsch / R. W. Schmitz / B. Worthington / J. E. Buikstra / A. Druzhkova / A. S. Graphodatsky / N. D. Ovodov / N. Wahlberg / A. H. Freedman / R. M. Schweizer / K. P. Koepfli / J. A. Leonard / M. Meyer / J. Krause / S. Pääbo / R. E. Green / R. K. Wayne, Complete mitochondrial genomes of ancient canids suggest a European origin of domestic dogs. Science 342, 2013, 871-874.

Thomas 2004: N. Thomas, The Early Mycenaean Lion Up To Date. In A. Chapin (ed.), ΧΑΡΙΣ. Essays in Honor of Sara A. Immerwahr. Hesperia Supplement 33, 2004, 161-206.

Thomas 2014: N. R. Thomas, A Lion's Eye View of the Greek Bronze Age. In: G. Touchais / R. Laffineur / F. Rougemont (eds), Physis: L'environnement naturel et la relation homme-milieu dans le monde égéen protohistorique (Leuven-Liege 2014) 375-389.

Thomas 2005: R. Thomas, Perceptions versus reality: changing attitudes towards pets in medieval and post-medieval England. In: A. Plukowski (ed.), Just Skin and Bones? New Perspectives on Human-Animal Relations in the Historic Past (Oxford 2005) 95-105.

Tilley 1996: C. Tilley, An Ethnographic of the Neolithic: Early Prehistoric Societies in Southern Scandinavia (Cambridge 1996).

Trolle-Lassen 1987: T. Trolle-Lassen, Human exploitation of fur animals in Mesolithic Denmark — a case study. Archaeozoologia 1, 1987, 85-102.

Turner 1992: A. Turner, Large carnivores and earliest European hominids: changing determinants of resource availability during the Lower and Middle Pleistocene. Journal of Human Evolution 22, 1992, 109-126.

Turner/Bateson 2014: D. C. Turner / P. P. G. Bateson (eds), The domestic cat: the biology of its behaviour (Cambridge 2014).

Turner/Waiblinger/Fehlbaum 2014: D. C. Turner / E. Waiblinger / B. Fehlbaum, Cultural differences in human–cat relations. In: Turner/Bateson 2014, 101-112.

U.S. Department of Agriculture 2016: U.S. Department of Agriculture, National Nutrient Database. Online in Internet: URL: http://ndb.nal.usda.gov/ (Stand 01.02.2016).

Vang Petersen 2013: P. Vang Petersen, Mesolithic Dogs. In: O. Grimm / U. Schmölcke (eds), Hunting in Northern Europe until 1500 AD. Old traditions and regional developments, continental sources and continental influences. Schriften des Archäologischen Landesmuseums: Ergänzungsreihe 7 (Neumünster 2013) 147-162.

Van Neer et al. 2014: W. Van Neer / V. Linseele / R. Friedman / B. De Cupere, More evidence for cat taming at the Predynastic elite cemetery of Hierakonpolis (Upper Egypt). Journal of Archaeological Science 45, 2014, 103-111.

Vigne et al. 2004: J. D. Vigne / J. Guilaine / K. Debue / L. Haye / P. Gérard, Early taming of the cat in Cyprus. Science 304/5668, 2004, 259-259.

Vilà/Maldonado/Wayne 1999: C. Vilà / J. E. Maldonado / R. K. Wayne, Phylogenetic relationships, evolution and genetic diversity of the domestic dog. Journal of Heredity 90, 1999, 71-77.

Vilà et al. 1997: C. Vilà / P. Savolainen / J. E. Maldonado / I. R. Amorim / J. E. Rice / R. L. Honeycutt / K. A. Crandall / J. Lundeberg / R. K. Wayne, Multiple and ancient origins of the domestic dog. Science 276, 1997, 1644-1648.

Vörös 1990: I. Vörös, Kutyaáldozatok és temetkezések a középkori Magyarországon I / Dog Sacrifices and Burials in Medieval Hungary I. Folia Archaeologica 42, 1990, 117-145.

Vretemark 2013: M. Vretemark, The Vendel Period royal follower's grave at Swedish Rickeby as starting point for reflections about falconry in northern Europe. In: O. Grimm / U. Schmölcke (eds), Hunting in Northern Europe until 1500 AD. Old traditions and regional developments, continental sources and continental influences. Schriften des Archäologischen Landesmuseums: Ergänzungsreihe 7 (Neumünster 2013) 379-386.

Wamers 2009: E. Wamers, Von Bären und Männern. Berserker, Bärenkämpfer und Bärenführer im frühen Mittelalter. Zeitschrift für die Archäologie des Mittelalter 37, 2009, 1-46.

2015. E. Wamers, Bärenkult und Schamanenzauber: Rituale früher Jäger (Regensburg 2015).

Wang et al. 2016: G. D. Wang / W. Zhai / H. C. Yang / L. Wang / L. Zhong / Y. H. Liu / R. X. Fan / T. T. Yin / C. L. Zhu / A. D. Poyarkov / D. M. Irwin / M. K. Hytönen / H. Lohi / C. I. Wu / P. Savolainen / Y. P. Zhang, Out of southern East Asia: the natural history of domestic dogs across the world. Cell Research 26, 2016, 21-33.

Wayne/Ostrander 1999: R. K. Wayne / E. A. Ostrander, Origin, genetic diversity, and genome structure of the domestic dog. BioEssays 21, 1999, 247-257.

Wenzel 2009: S. Wenzel, Behausungen im späten Jungpaläolithikum und Mesolithikum Nord-, Mittel- und Westeuropas. Monographien des RGZM 81 (Mainz 2009).

Wiedemann 2002: R. E. Wiedemann, Die Tiere der Gesellschaft (Konstanz 2002).

Wigh 2001: B. Wigh, Animal Husbandry in the Viking Age Town of Birka and its Hinterland (Stockholm 2001).

Yannouli 2003: E. Yannouli, Non-domestic carnivores in Greek prehistory: a review. In: E. Kotjabopoulou / Y. Hamilakis / P. Halstead / C. Gamble (eds), Zooarchaeology in Greece: Recent Advances (London 2003) 175-192.

Zimen 1978: E. Zimen, Der Wolf. Mythos und Verhalten (Wien, München 1978).

SUMMARY / ZUSAMMENFASSUNG

Humans and their Relationship to Large Carnivores in Central Europe from Foragers to Modern Times: A Survey

Carnivores were always exploited for their pelts and their meat, even if the latter function might have been in many cases a secondary product only. In particular the human relation to wolves and dogs far exceeds the level of pure hunting objects and dietary source. Since Mesolithic times dogs were emotional objects for their owners with an semi-human status. Through all times and on most continents dogs were regular buried, partly just to honour them, partly because the dog had to be its owner's companion and helper also in the afterlife.

Cats, although domesticated since four millennia and living closely to humans in Central Europe since 1000-2000 years, have not developed such a strong connection to humans and vice versa. In contrast to dogs, graves of or with cats are scarce, and up to modern times the human relation to cats was characterised by reservation. This is caused by the biology of the cat's behaviour and led for millennia to a more utilitarian function of this mouse-hunting species.

The wildcat, but also the larger species wolf and bear were no economically relevant objects of human hunting for many thousands of years after the end of the Ice Age. Only in the 1st millennium AD bear hunting became more common, but potentially often limited to special purposes. In several regions a connection either to burial costums or to hunting, performed by the social elite, can be recorded.

Remains and even indirect references of wolves in human context are particularly rare from Central European Stone to Iron Age. Without question the fur of wolves was used, but it was of minor relevance in comparison to the pelt of other species such as beaver or fox.

In the course of Christianisation, both bears and wolves were demonized and hunted to extinction in large parts of Europe. Recently, a comeback is to observe, accompanied by social discussions. However, the increasing population of wolf, bear, and lynx – partly due to wildlife conservation or restoration programs – could be a reflection of a slowly changing attitude towards these species in the Central European human society.

Der Mensch und seine Beziehung zu Großraubtieren in Mitteleuropa von der Steinzeit bis zur Neuzeit: ein Überblick

Karnivoren wurden schon immer wegen ihrer Felle und ihres Fleisches gejagt, auch wenn letzteres in vielen Fällen nur ein Nebenprodukt gewesen sein mag. Insbesondere die Beziehung des Menschen zu Wölfen und Hunden geht weit über die Ebene des reinen Jagdobjekts hinaus. Seit dem Mesolithikum waren Hunde für ihre Besitzer emotionale Objekte mit einem halbmenschlichen Status. Zu allen Zeiten und auf den meisten Kontinenten wurden Hunde regelmäßig begraben, teils um sie zu ehren, teils weil der Hund seinem Besitzer auch im Jenseits als Begleiter und Helfer dienen sollte.

Katzen, obwohl seit vier Jahrtausenden domestiziert und in Mitteleuropa seit 1000-2000 Jahren eng mit dem Menschen zusammenlebend, haben keine so starke Bindung zum Menschen entwickelt und umgekehrt. Im Gegensatz zu Hunden sind Gräber von oder mit Katzen in der Archäologie rar, und bis in die Neuzeit war das Verhältnis des Menschen zu Katzen von Vorbehalten geprägt. Dies ist in der Biologie des Katzenverhaltens begründet und führte über Jahrtausende zu einer eher utilitaristischen Funktion dieser mäusejagenden Spezies.

Die Wildkatze, aber auch die größeren Arten Wolf und Bär waren viele Jahrtausende nach dem Ende der Eiszeit keine wirtschaftlich relevanten Jagdobjekte. Erst im 1. Jahrtausend n. Chr. wurde die Bärenjagd häufiger, aber oft auf spezielle Zwecke beschränkt. So lässt sich in mehreren Regionen ein Zusammenhang entweder mit Bestattungssitten oder mit der von der gesellschaftlichen Elite ausgeübten Jagd feststellen.

Überreste und sogar indirekte Hinweise auf Wölfe im menschlichen Kontext sind von der mitteleuropäischen Stein- bis zur Eisenzeit besonders selten. Zweifellos wurde das Fell von Wölfen verwendet, doch war es im Vergleich zum Fell anderer Arten wie Biber oder Fuchs von geringerer Bedeutung.

Im Zuge der Christianisierung wurden sowohl Bären als auch Wölfe dämonisiert und in weiten Teilen Europas bis zur Ausrottung gejagt. In jüngster Zeit ist ein Comeback zu beobachten, begleitet von gesellschaftlichen Diskussionen. Die zunehmende von Wolf, Bär und Luchs – zum Teil aufgrund von Artenschutz- oder Wiederansiedlungsprogrammen – könnte jedoch auch Ausdruck einer sich langsam ändernden Einstellung der mitteleuropäischen Gesellschaft gegenüber diesen Tierarten sein.

CLARA STEFEN

THE EUROPEAN WILDCAT *FELIS SILVESTRIS* – A BRIEF REVIEW OF STATUS AND RESEARCH IN GERMANY

The European wildcat is a Palaearctic species with an initially wide distribution from the Iberian Peninsula to the Caucasus and Scotland (Piechocki 1990; Hemmer 1993; Heptner/Sludskij 1980). After the eradication of carnivores was enforced (e.g. von Hoberg 1687), especially in the 18th and 19th century, it was nearly extirpated at the beginning of the 20th century. Therefore, its distribution is discontinuous (Mitchell-Jones 1999; Aulagnier et al. 2009). It is generally assumed that relict populations have survived in the northeast and east of France, the south of Belgium, Luxembourg, the west and southwest of Germany and areas of the Harz Mountains, and in Eastern Europe, mainly in the Carpathian Mountains (Hemmer 1993, 1100). Within Germany wildcats survived in the Eifel, Hunsrück, Harz Mountains (de Leuw 1976) and probably Thuringia, the Taunus, Black Forest, Palatinate Forest and some areas in Hesse (Petzsch 1968; Röben 1974). Due to near extirpation, the wildcat is protected in Germany since 1934 in the Reichsjagdgesetz (Görner 2012). This – and a reduced hunting activities during and after World War II – was probably the main reason for the survival of the species in Germany (Röben 1974). Since 1992 it is also strongly protected in Europe under the European Habitat Directive where it is listed in appendices II and IV. These measures led to increasing population size and dispersal in recent years (i.e. Görner 2012; Krug et al. 2012; Sodeikat/Köglsperger 2012; Streif et al. 2012), but also a greater effort was made in detecting the species than before. Most dispersal occurred by natural population growth, but there was also a reintroduction programme in Bavaria where 580 wildcats were released from 1984 to 2008 (Worel 2009; Heinrich 1992). The current distribution range in Germany is quite substantial and increased over the decades (**fig. 1**). The national report in relation to the Habitat Directive (BFN 2013) gives a minimum of 914 and maximum of 934 individuals in Germany.

The domestic cat is considered to be the result of the domestication of the African wildcat (*Felis silvestris lybica*) about 9000 years ago in the villages of the Fertile Crescent in the Near East (Driscoll et al. 2007; 2009). Hard evidence from the early period of domestication (9500-4000 years ago) is lacking. It is only from approx. 4000 years ago that domestic cats are represented in ancient Egyptian art. However, there is evidence from an early agricultural village in China from about 5560-5280 years BP (Hu et al. 2013). A cat buried next to a human grave on Cyprus dating about 9500 years ago is considered to be the first evidence that humans kept cats as pets (Vigne et al. 2004). It is assumed that African wildcats followed humans because of the rich source of mice attached to their cereal storages. As an important pest control the cat obtained a special status in Egypt; by about 2900 years ago, it was the symbol of a god (Driscoll et al. 2009). Later the domestic cat was widely distributed by the Romans who also treasured it for keeping mice at bay about 2000 years ago (Clutton-Brock 1987; Daniels et al. 1998; Driscoll et al. 2009). The Romans also introduced them (probably in small numbers) to Central Europe but even more so during the Carolingian period about 1100 years ago (Knapp et al. 2002). Hamilton (1869), however, assumed that Egyptian cats were already introduced into Europe about 300-400 years BC. Cats were treasured because of their pest control qualities but in medieval times the tide turned and the image was ambivalent; in high medieval times the cat was seen as a symbol of the devil (see Plos 2015 for more information).

In: Nicholas J. Conard · Henning Hassmann · Kurt Felix Hillgruber · Jordi Serangeli · Thomas Terberger (eds), The *Homotherium* Finds from Schöningen 13II-4. Man and Big Cats of the Ice Age. Forschungen zur Urgeschichte aus dem Tagebau von Schöningen 4 (Mainz 2022). DOI: https://doi.org/10.11588/propylaeum.1006.c13528

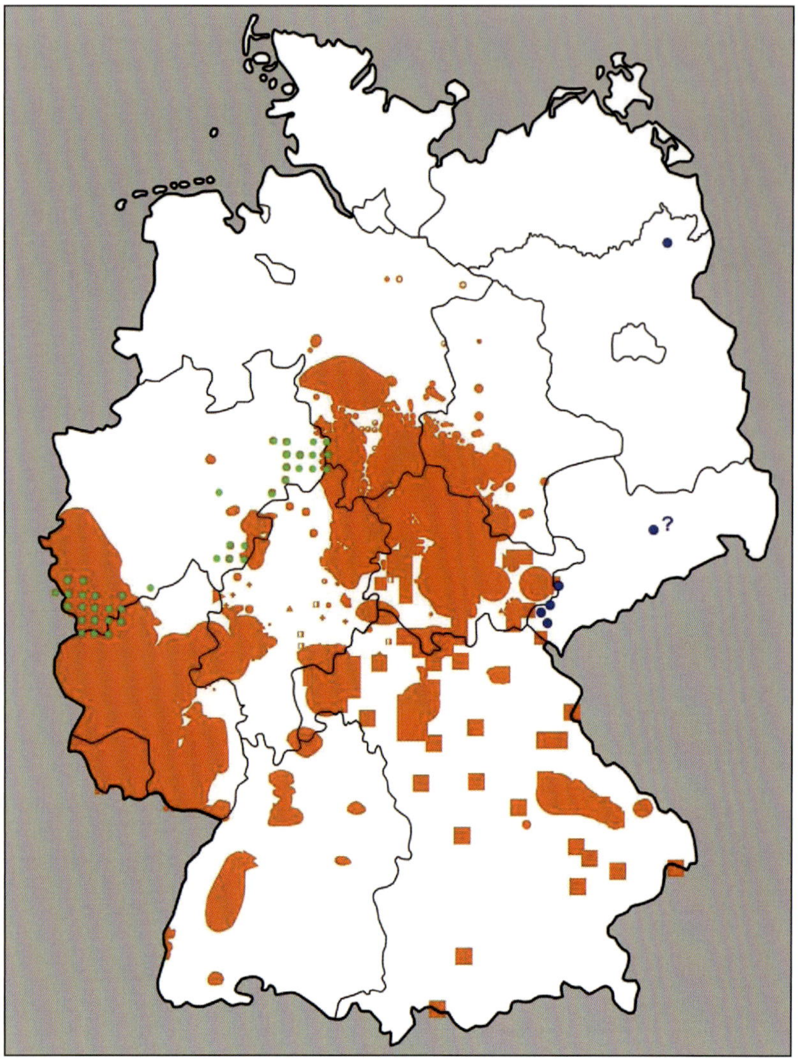

Fig. 1 Map indicating the distribution of wildcats in Germany derived from 28 maps published between 1957 and 2009 scaled to the same size (adapted after Stefen/Görner 2009; Stefen 2012). Included are maps from: Haltenorth 1957; Röben 1974; Jost 1978; Vogt 1985; Pflüger 1987; Piechocki 1987; 1989; 1990; Büttner/Worel 1990; Herrmann 1991; Piechocki/Möller 1991; Hossfeld et al. 1993; Raimer 1994; Stubbe/Stubbe 1994; Kock/Altmann 1999; Görner 2000; Raimer 2001; Knapp et al. 2002; Mölich/Klaus 2003; BUND 2004; Denk/Jung 2004; Frobel/Thein 2006; NABU 2007; Munlv NRW 2007; Pott-Dörfer/Dörfer 2007; Simon/Raimer 2007; Götz/Roth 2007; Stefen et al. 2009, and information according to Hucht-Ciorga (2011; in green), Stefen (2011; in blue) and new material in the Senckenberg Naturhistorische Sammlungen Dresden (2015; in black).

HISTORY, SYSTEMATICS AND TAXONOMY OF WILDCATS

The family Felidae probably originates from a predecessor that lived about 10-15 million years ago (Johnsen/O'Brien 1997). The first felid-like carnivores are known from the Oligocene about 35 million years ago, the genera of today arose during the Miocene. The most recent lineages within the family, the domestic cat lineage and the leopard lineages only separated about 6.2 to 6 million years ago (Johnson/O'Brien 1997; Johnson et al. 2006). Based on mtDNA the domestic cat lineages including *F. catus, F. silvestris, F. lybica, F. bieti* (these species are also sometimes summarized as subspecies of *F. silvestris* [Driscoll et al. 2009]), *F. margarita* and the older but associated *F. nigripes* and *F. chaus* started to separate 6.2 (Johnson et al. 2006) or 6 million years ago (Johnson/O'Brien 1997).

According to pelage colouring, size and robustness, three groups of wildcats are distinguished: the *silvestris*-group (European wildcats), the *lybica*-group (African wildcats) and the *ornata*-group (Indian wildcats; Haltenorth 1953; Weigel 1972; Kitchener 1991). The differentiation of wildcats probably occurred during the Pleistocene (Hemmer 1984), *lybica* and *silvestris* separated approximately 20 000 years ago during the cool periods of the Late Pleistocene (Hemmer 1984; 1993; Randi/Ragni 1991). The distribution of Late Pleistocene and Early Holocene finds of wildcats in Europe have been summarized by Sommer/Benecke (2006).

characteristic	wildcats	domestic cats
cranial volume	32.5-50 cm³ if between 32 and 35 cm³, the cranial index has to be calculated and used	2-35 cm³
cranial index (total skull length: cranial volume)	<2.75	>2.75
intestine length (measured from end of stomach lying flat or hanging)	120-170 cm, males 110-150 cm, females	165-254 cm, males 155-220 cm, females
intestine index (intestine length: head-body length)	2.04-3.17	3.20-4.84
skull: glabella, nasal bones	no depression at the suture between nasal bones and frontal	a slight depression at the suture between nasal bones and frontal, and nasals extending further into frontal
mandible: angular process (but variable and changing during ontogeny!)	angular process usually extending further caudally; mandible stands if placed on coronoid, condyle and angular processes	angular process usually not extending as far caudally as coronoid and condyle; mandible paced on coronoid, condyle and angular processes is instable and falls
neck pelage pattern	3-5 clearly separated stripes	stripes usually not as cleary separated
basic underlying fur colour	brownish, ochreous	greyish, greater variability in shades of grey
basic pelage pattern	indistinct, vague stripes	tabby, stripes more distinct and usually stronger in colour

Tab. 1 Distinctive characteristics to distinguish between wildcats and domestic cats (in wildcat colouring) in accordance to Schauenberg (1969; 1977), Piechocki (1990), Piechocki/Möller (1983).

DIFFERENTIATION FROM DOMESTIC CATS – MORPHOLOGY AND MOLECULAR DATA

In the following paragraphs the European wildcat is solely referred to as wildcat. One of the long debated issues that is of importance in species conservation is the differentiation of the cat forms and particularly the separation of domestic cat and wildcat. Both species are close enough that they can hybridize. There are many papers dealing with the characteristics of wildcats and their differentiation from domestic cats, mainly based on morphology or molecular data (i. e. Schauenberg 1969; 1971; 1977; Piechocki 1990; Daniels et al. 1998; Beaumont et al. 2001; Reig et al. 2001; Kitchener et al. 2005; Müller 2005; Devillard et al. 2014). Several studies deal with the skull morphology of wildcats (e. g. Sládek et al. 1971; 1972; Kratochvíl/Kratochvíl 1970; Kratochvíl 1973; French et al. 1988; Yamaguchi et al. 2004; Krüger et al. 2009), some also describe the craniometric variability between and within different populations (Krüger et al. 2009; Stefen/Heidecke 2011; Stefen 2012a). Other studies deal with the outer phenotype, the pelage pattern and colouring (Eckstein 1919; Vogt 1985; Müller 2011a; 2011b; Ragni/Possenti 1996; Reher/Stefen submitted). Wildcats and domestic cats can be distinguished morphologically on the basis of a few characteristics (**tab. 1; figs 2-5**) but hybrids cannot really be detected.

In outer morphology the most striking and distinctive characteristic of the wildcat is the blunt, relatively thick tail (due to long and dense hairs) with marked, separated dark (black) bands and a black tip. Also on the back of the head and neck there are well distinguished stripes (often not as well developed in domestic cats), continuing on the forehead, two separated stripes above the shoulder, and a black dorsal stripe end-

Fig. 2 Old illustration of a wildcat by Blasius 1857 indicating some characteristic features: blunt, thick tail with distinctive black bands and black tip, black eel back, two stripes above the shoulders, distinct stripes on head and neck. Only the stripes at the back are too distinct for a typical German wildcat. – (After Blasius 1857).

ing at the base of the tail. The overall base colour is brownish or ochreous and the hairs, particularly the guard hairs, are very long and fine (often described as silky). But there is a substantial overlap in hair length to domestic cats. The overall pelage pattern of wildcats are faint stripes, and usually there is a white spot on the throat. Anatomically the most distinctive characters between wildcats and domestic cats are cranial volume, cranial index, intestine length and intestine index (see **tab. 1**). There are other differences listed, which are, however, not as distinctive as those given characteristics. Krüger et al. (2009) consider the length of the lower canines, width between the frontal-parietal suture at the smallest position of the orbita, length of lower p4, length between lower p2 and m1, coronoid height, rostrum width at canines as distinctive.
In recent decades molecular analyses are used to differentiate between domestic and wildcats; a regularly used technique is genotyping using several genetic markers (usually more than ten microsatellite loci – loci in the nuclear genome) (i. e. Daniels et al. 2001; Lecis et al. 2006; Oliveira et al. 2007; Hertwig et al. 2009; Eckert et al. 2010; Driscoll et al. 2011; Hartmann et al. 2013; Steyer et al. 2013). This method needs a large sample size and a good database of well defined species as reference sample as it works with probabilities of group membership and has been criticised (Gehle 2012; Gehle/Herzog 2012). Nussberger et al. (2013, 2014) developed another method using single nucleotide polymorphism (SNP) markers to differentiate both cat forms and to detect introgression, thus hybridization, and is now widely accepted.

REPRODUCTION, DEVELOPMENT OF JUVENILES, AGE AND DEATH RATES

The wildcat is a (mainly) monoestric species with the main runt in winter, November to early March with the main period in February and March (see Stefen/Görner 2009). After a gestation period of 66 days on average about 2-4 young are born (Hemmer 1993). Some juveniles have also been recorded in other seasons so it is assumed that a second or even third period of runt is possible (Piechocki/Möller 1983). Weaning usually occurs by the end of the fourth month »when survival rate of cubs was 20 %« (Götz et al. 2009). The main mortality cause is predation by mustelids (Götz/Roth 2007). Probably free ranging dogs might also be listed here. The role of the increasing number of raccoons (*Procyon lotor*), known to take eggs, nestlings

Fig. 3 Comparison of pelage characteristics illustrated for Scottish wildcats (**A**) and putative domestic cats (**B**): **1** white on chin. – **2** stripes on cheek. – **3** dark spots on ventral side. – **4** white on paw. – **5** white on flank. – **6** white on back. – ***7*** extent of dorsal line. – ***8*** shape of tail tip. – **9** colour of tail tip. – ***10*** distinctness of tail bands. – **11** alignment of tail bands. – **12** stripes on hind leg. – **13** bands encircling foreleg. – **14** tabby coat patterns. – ***15*** broken stripes on flanks & hindquarters. – **16** stripes on body. – ***17*** spots on flanks & hindquarters. – ***18*** stripes on nape. – ***19*** stripes on shoulder. – **20** colour of the back of the ears. – Properties best suited for discrimination of **A** and **B** are in italics and in black boxes. – (After Kitchener et al. 2005).

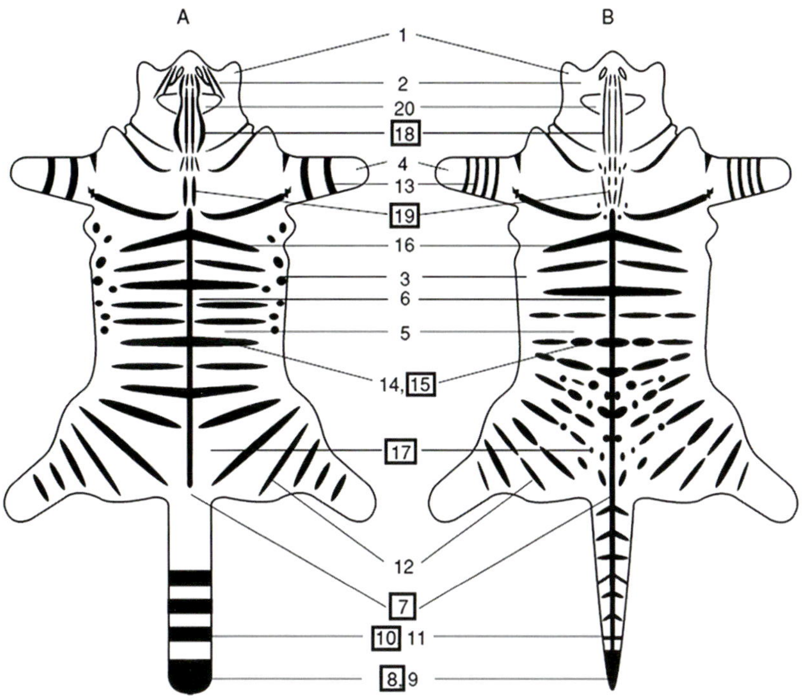

and disturb breeding birds, on wildcats raising pups is unknown. Reaching adulthood, the wildcat does not have real enemies, though it might be killed by lynx (Raimer 2006), considered its foremost enemy for a long time (Haltenorth 1957), or potentially by large birds of prey (Schauenberg 1970).
The life expectancy has been stated to reach 16 years (Piechocki/Möller 1983), but Piechocki/Stiefel (1988) assumed that a life expectancy of 12-15 years in nature is too high. Büttner (1994) noted that the life expectancy has to be corrected to 6 years only. Recently, however, a 12-year-old female was detected in Thuringia (Krüger et al. 2015).

Fig. 4 Wildcat in the Senckenberg Naturhistorische Sammlungen Dresden (no data and no collection number) displaying some typical pelage characteristics and the thick banded tail. – (Photo C. Stefen).

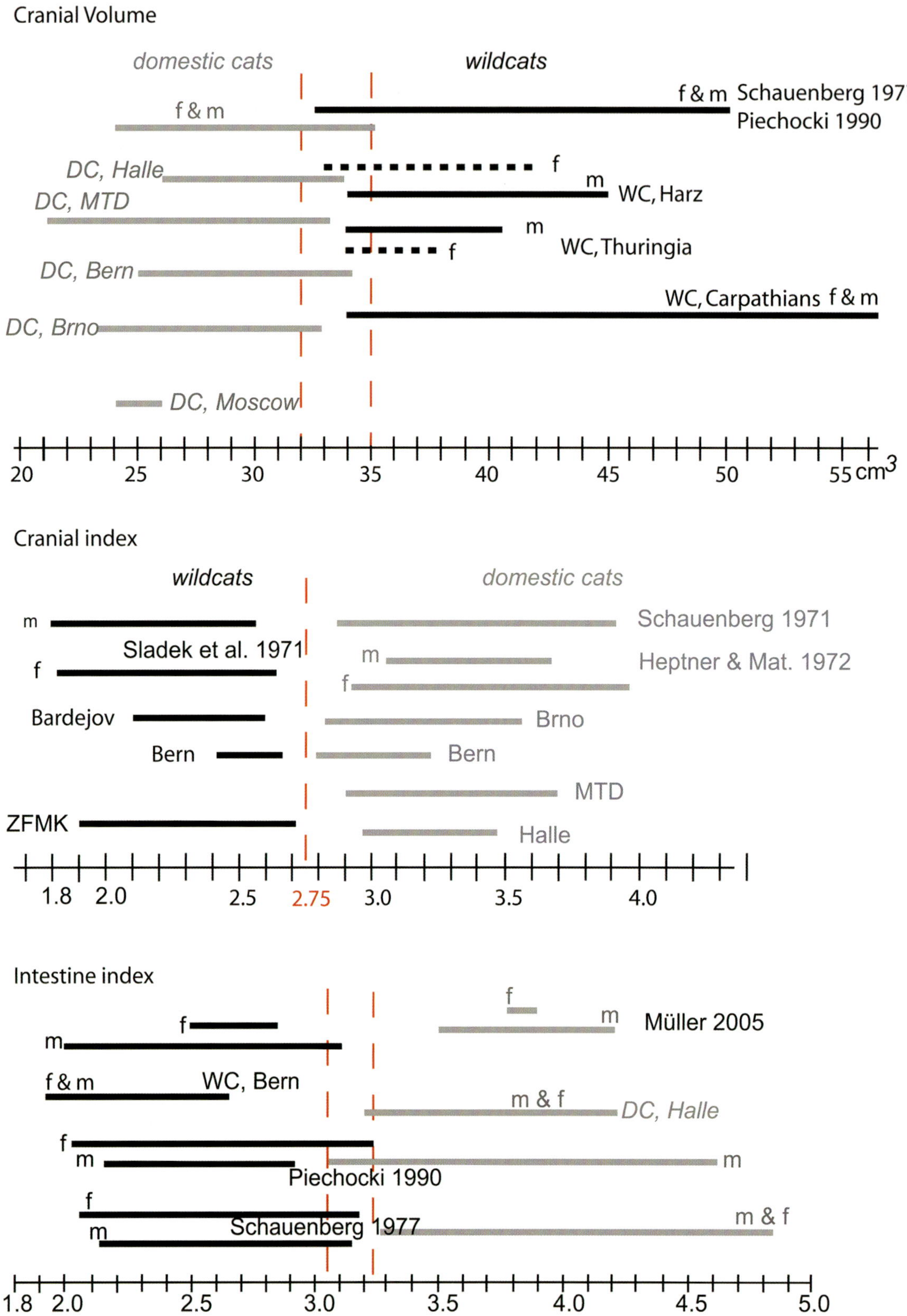

Fig. 5 Data of cranial volume, cranial index and intestine index for wildcats (WC) and domestic cats (DC) according from literature (REFS) or from own measurements in different collections: Bardejov – Museum Bardejov, Slovakia, SMB; Bern – Naturhistorisches Museum der Burgergemeinde Bern, Switzerland, NMBE; Brno – Institute of Vertebrate Biology, Academy of Sciences of the Czech Republic; Halle – Zoological Institute of the Universtity Halle-Wittenberg, now part of the central repository of natural history collections, Germany; MTD – Senckenberg Naturhistorische Sammlungen Dresden, Museum für Tierkunde, Germany; and ZFMK – Zoologisches Forschungsmuseum Alexander König, Bonn, Germany; f – females; m – males.

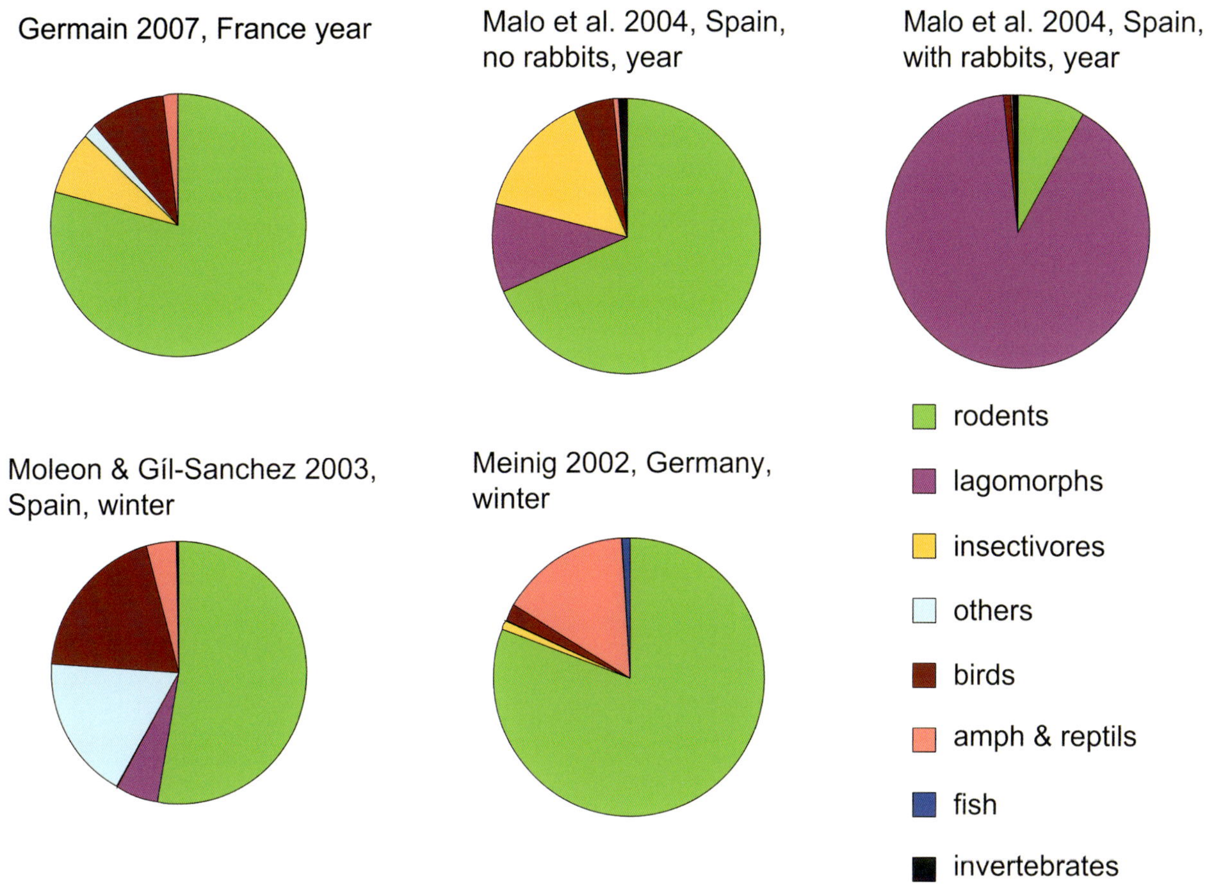

Fig. 6 Proportions of food sources consumed by wildcats from some studies (REFS) indicating the variability in intake of taxonomic groups. Given is the study, the country and season of the study. Malo et al. (2004) compared two regions, one with rabbits and one without. Amph – amphibia.

DIET

Cats eat mice and this is true for wildcats as well. But of course, the diet of both cat forms is much more diverse (**fig. 6**) as it has been shown in different studies. The main prey animals of wildcats in Central Europe are common voles (*Microtus arvalis*) but also other small mammals, like shrews, Muridae, Arvicolidae, Cricetidae, Lagomorpha, birds, amphibians and reptiles, fish, invertebrates, many insects and arthropods, plant matter, and even artiodactyls as carrion (i.e. summarized by Haltenorth 1957; Sládek 1973a; Stefen 2012b). The wildcat's diet depends on regional occurrence but »depends more on the factor of availability than that of abundance of prey« (Sládek 1973b, 143). The diet changes over the year (i.e. Sládek 1973b), and can vary between regions. In Spain, rabbits or hares are preferred (Malo et al. 2004; Moleón/Gil-Sánchez 2003). Therefore, wildcats have been termed facultative specialists but might even behave as a generalist species (Malo et al. 2004). Hansen (2014) documented a wildcat killing a hare in Germany and for example Heidemann (1973), Splitter (1978) or von Borkenhagen (1979) documented lagomorphs as food of domestic cats. According to historic literature the common hamster was also part of the wildcat's diet and a change in prey species with the decreased availability of some prey species for the wildcat has been hypothesized (Stefen 2012b). In general it can be assumed that the overall abundance of small mammals in fields decreased over the last century due to the use of rodenticides.

Tab. 2 Habitat requisites needed or used by wildcats mentioned in the analysed literature (from 100 publications with reference to the habitat, Stefen/Görner 2009).

	requisite mentioned	number of mentions
1	rocks, cracks	31
2	hollows in trees	28
3	dens of badger or fox	23
4	coppice, undergrowth	20
5	roots of fallen trees	13
6	depression in the ground	9
7	empty sheds, hunting stands	8
8	empty nests of birds of prey	3
9	tree studs	4
10	strong vertical branches, crotches	1
	old bunkers in the Eifel (Western Germany)	5
11	heaps of brushwood, or cut wood	11
12	low branches	5
13	hay racks	2
14	young plantations	1

ECOLOGY AND HABITAT

To briefly summarize the ecology and habitat preferences of wildcats is difficult and for more details see Stefen/Görner (2009). Overall, the wildcat can be considered a species with few important basic requirements, primarely cover, food and dry spots without drafts for raising their young. In addition, modern wildcats have been shown to use different kinds of habitats utilising varied requirements (**tab. 2**).

The older descriptions of preferred habitats like continuous old growth woodlands in mainly the rugged or rocky areas of low mountain ranges probably result from the fact that such areas were not easily accessible to humans and thus were ideal refuge areas for wildcats. Woodlands and forests are dynamic structures in time due to natural processes and human activity. In the middle of the 18th century, Central European forests were strongly overused by humans: traditional use as wood pasture, the collection of forest litter and firewood but also the need of wood for rafting of timber, metallurgical processes, heating, building etc. increased the demand of wood. Hunting and the collection of wild berries also occurred in woods and forests. The natural rejuvenation of beech and silver fir was impossible. At the end of the 18th century, at least in Thuringia/D did not have continuous closed woodlands any more (Witticke 2015). The wildcat had to adapt to this modified habitat. The main prey species changed to *Microtus arvalis*, which lives in open grasslands and agricultural areas, but not in woods or forests. The wildcats increased use of open areas is shown by telemetric studies (i. e. Jerosch/Götz 2011) and the most typical habitat today is a mosaic landscape (Lozano et al. 2003) of wood and open grasslands, the edges or contact zones between the habitats are apparently important as well as water courses and the accompanying vegetation (Heinrich 1992).

As the observed use of anthropogenic structures indicates, wildcats can tolerate human activities (Vogt 1985) and their records increase close to and in villages, particularly in very cold winters (Piechocki 1990). This trend is also observed in Russia (Heptner/Sludkii 1980), Scotland (Scott et al. 1993) and France (Artois 1985).

HOME RANGES

As a rule, the home ranges of wildcats range from 1000 to 2000 ha (summarised in Stefen/Görner 2009). The home ranges of males are usually larger than those of females and overlap those of several females (i. e. Hötzel et al. 2007; Krug et al. 2012). The size of the home ranges varies with the season, and with females decreases when young are being reared. The ranges of two individuals of the same sex usually do not overlap (Hupe et al. 2004), but there are exceptions for females (Götz/Roth 2007; Hötzel et al. 2007) as well as for males (Wittmer 2001, 366; Mölich/Klaus 2003, 127; Götz/Roth 2007).

ENDANGERMENT AND PROTECTION

Most authors agree that the destruction or fragmentation of the wildcat habitat and the hybridization with domestic cats are the two main factors endangering wildcats (i. e. Stahl/Artois 1991; Hubbard et al. 1992; Daniels/Corbett 2003; Lozano et al. 2007). Fragmentation of habitats is often used very figuratively in conservation. It is assumed that wildcats cannot cross structures that are considered as barriers by humans. These are mainly roads or railway lines, which can only be crossed in dangerous circumstances, thus most cats are killed by traffic (Klaus et al. 2012). The wildlife protection fences at these structures reduce traffic casualties but turn them into insurmountable barriers. Therefore, green bridges, over-/underpasses or valley bridges are important for wildcats.

FUTURE PERSPECTIVE

In recent years wildcats have been detected in more and more regions, particularly with the use of lure sticks to collect hairs and the subsequent genetic analysis (Hupe/Simon 2007; Steyer et al. 2013). The increasing population and dispersal is also indicated by the increasing number of wildcats casualties in traffic.

The (increasing?) hybridization with domestic cats (which is debated as indicated above) has to be reduced if the protection of the »genetic« status of the wildcat is to be given priority. The older practice of killing domestic cats outside of human settlements is hardly implemented anymore. Assuming that not all domestic cats are sterilized, it might be hypothesized that the hybridization potential could increase. On the other hand, wildcat populations also increase, which might make it easier to find adequate partners and the hybridization potential remains the same or even decreases. But increasing numbers of free ranging or feral domestic cats might have another negative impact on wildcats: As their prey preferences are very similar (i. e. Biró et al. 2005; Germain et al. 2009; Stefen 2012b), more domestic cats may limit the food resources of wildcats, particularly in periods with extreme weather reducing small mammal abundance. In this context the still common practice of using rodenticides in years with an estimated vole (*Microtus arvalis*) gradation (year of mass repodution) has to be questioned.

Another possible new endangerment of wildcats might be climate change. Within the wildcats the European wildcat has adapted to cooler climates and thus might be more susceptible to an increasingly warmer climate in Central Europe, but up to now no change in size could be attributed to climate change (Stefen 2015). However, climate change might also induce a change in the distribution of parasites and pathogens, or have impact on a species physiology and immunology.

ACKNOWLEDGEMENTS

I heartily thank the organizers of the very interesting workshop »The *Homotherium* finds from Schöningen 13II-4 and big cats of the Ice Age« for their invitation.

REFERENCES

Artois 1985: M. Artois, Utilisation de l'espace et du temps chez le renard (*Vulpes vulpes*) et le chat forestier (*Felis silvetris*) en Lorraine. Gibier Faune Sauvage 3, 1985, 33-57.

Aulagnier et al. 2009: S. Aulagnier / P. Haffner / A. J. Mitchell-Jones / F. M. Moutou / J. Zima, Die Säugetiere Europas, Nordafrikas und Vorderasiens: Der Bestimmungsführer (Bern, Stuttgart, Wien 2009).

Beaumont et al. 2001: M. Beaumont / E. M. Barrat / D. Gottelli / A. C. Kitchener / M. J. Daniels / J. K. Pritchards / M. W. Bruford, Genetic diversity and introgression in the Scottish wildcat. Molecular Ecology 10, 2001, 319-336.

BFN 2013: Vollständige Berichtsdaten zum Nationalen FFH-Bericht 2013. www.bfn.de/themen/natura-2000/berichte-monitoring/nationaler-ffh-bericht/2013-ffh-bericht/berichtdaten-2013.html (14.9.2020).

Biró et al. 2005: Z. Biró / J. Lanszki / L. Szemethy / M. Heltai / E. Randi, Feeding habits of feral domestic cats (*Felis catus*), wild cats (*Felis silvestris*) and their hybrids: trophic niche overlap among cat groups in Hungary. Journal of Zoology 266, 2005, 187-196.

Blasius 1857: J. H. Blasius, Naturgeschichte der Säugethiere Deutschlands und der angrenzenden Länder von Mitteleuropa (Braunschweig 1857).

von Borkenhagen 1979: P. von Borkenhagen, Zur Nahrungsökologie streunender Hauskatzen (*Felis silvestris* f. *catus* L., 1758) aus dem Stadtbereich Kiel. Zeitschrift für Säugetierkunde 44, 1979, 375-383.

BUND 2004: Ein Rettungsnetz für die Wildkatze, BUND Thüringen, BUND Hessen, BUND Bayern.

Büttner 1994: K. Büttner, 10 Jahre Wiedereinbürgerung der Europäischen Wildkatze *Felis silvestris silvestris* Schreber 1977 in Bayern (1984-1993). Waldhygiene 20, 1994, 137-145.

Büttner/Worel 1990: K. Büttner / G. Worel, Wiedereinbürgerung der europäischen Wildkatze in Bayern – ein Projekt des Bundes Naturschutz in Bayern. Waldhygiene 18, 1990, 169-176.

Clutton-Brock 1987: J. Clutton-Brock, A Natural History of Domesticated Mammals (Cambridge, London 1987).

Daniels/Corbett 2003: M. J. Daniels / L. Corbett, Redefining introgressed protected mammals: when is a wildcat a wild cat and a dingo a wild dog? Wildlife Research 30, 2003, 213-218.

Daniels et al. 1998: M. J. Daniels / D. Balharry / D. Hirst / A. C. Kitchener / R. J. Aspinall, Morphological and pelage characteristics of wild living cats in Scotland: implications for defining the »wildcat«. Journal of Zoology 244, 1998, 231-247.

2001: M. J. Daniels / M. A. Beaumont / P. J. Johnson / D. Balharry / D. W. Macdonald / E. Barratt, Ecology and genetics of wild-living cats in the north-east of Scotland and the implications for the conservation of the wildcat. Journal of Applied Ecology 38, 2001, 146-161.

Denk/Jung 2004: M. Denk / J. Jung, Die Situation der Wildkatze. In: Hessen, Hessisches Ministerium für Umwelt, ländlichen Raum und Verbraucherschutz (ed.), Reihe Natura 2000 (Wiesbaden 2004).

Devillard et al. 2014: S. Devillard / T. Jombart / F. Léger / D. Pontier / L. Say / S. Ruette, How reliable are morphological and anatomical characters to distinguish European wildcats, domestic cats and their hybrids in France? Journal of Zoological Systematics and Evolutionary Research 52, 2014, 154-162.

Driscoll et al. 2007: C. A. Driscoll / M. Menotti-Raymond / A. L. Roca / K. Hupe / W. E. Johnson / E. Geffen / E. H. Harley / M. Delibes / D. Pontier / A. C. Kitchener / N. Yamaguchi / S. J. O'Brien / D. W. Macdonald, The Near Eastern Origin of Cat Domestication. Science 317/5837, 2007, 519-523.

2009: C. A. Driscoll / J. Clutton-Brock / A. C. Kitchener / S. J. O'Brien, The Evolution of House cats. Scientific American 300/6, 2009, 56-63.

2011: C. Driscoll / N. Yamaguchi / S. J. O'Brien / D. W. Macdonald, A Suite of Genetic Markers Useful in Assessing Wildcat (*Felis silvestris* ssp.)-Domestic Cat (*Felis silvestris catus*) Admixture. Journal of Heredity 102 (Suppl. 1), 2011, 87-90.

Eckert et al. 2010: I. Eckert / F. Suchentrunk / G. Markov / G. B. Hartel, Genetic diversity and integrity of German wildcat (*Felis silvestris*) populations as revealed by microsatellites, allozymes, and mitochondrial DNA sequences. Mammalian Biololgy 75/2, 2010, 160-174.

Eckstein 1919: K. Eckstein, Beiträge zur genaueren Kenntnis der Wildkatze. Archiv für Naturgeschichte 85, Abt. A, 12, 1919, 1-59.

French/Corbett/Easterbee 1988: D. D. French / L. K. Corbett / N. Easterbee, Morphological discriminants of Scottish wildcats *Felis silvestris*, domestic cats *F. catus* and their hybrids. Journal of Zoology 214/2, 1988, 235-259.

Frobel/Thein 2006: K. Frobel / J. Thein, Die Wildkatzen in Bayern: Stand und Ausblick. In: Naturschutz-Akademie Hessen, Bund für Umwelt und Naturschutz Deutschland, Institut für Tierökologie und Naturbildung (ed.), Kleine Katzen – große Räume: ein Rettungsnetz für die Wildkatze. Naturschutz-Akademie Hessen (Wetzlar): Akademie-Berichte 5 (Wetzlar 2006) 109-114.

Gehle 2012: T. Gehle, Wie wild ist die Wildkatze? Grenzen der Diagnostik. Rheinisch-Westfälischer Jäger 9, 2012, 16-17.

Gehle/Herzog 2012: T. Gehle / S. Herzog, Sinn und Unsinn einer Differenzierung von Wild- und Hauskatzen mit Hilfe genetischer Marker. Säugetierkundliche Informationen 8/45, 2012, 329-336.

Germain/Ruette/Poulle 2009: E. Germain / S. Ruette / M.-L. Poulle, Likeness between the food habits of European wildcats, domestic cats and their habrids in France. Mammalian Biology 74/5, 2009, 412-417.

Görner 2000: M. Görner, Zum Vorkommen der Wildkatze (*Felis silvestris*) in Thüringen von 1800 bis 2000. Artenschutzreport 10, 2000, 54-60.

2012: M. Görner, Wildkatzen (*Felis silvestris*) in Thüringen aus aktueller Sicht von Jagd, Naturschutz und Wissenschaft. Säugetierkundliche Informationen 8/45, 2012, 387-403.

Götz/Roth 2007: M. Götz / M. Roth, Verbreitung der Wildkatze (*Felis s. silvestris*) in Sachsen-Anhalt und ihre Aktionsräume im Südharz. Beiträge zur Jagd- und Wildforschung 32, 2007, 437-447.

Götz/Jerosch/Roth 2009: M. Götz / S. Jerosch / M. Roth, Reproductive parameters of European wildcat and the importance of dead wood structures. In: 83rd Annual Meeting of the German Society of Mammalogy Dresden, 13 to 17 September 2009 Mammalian Biology: Special Issue to Vol. 74, 2009, 11.

Haltenorth 1953: Th. Haltenorth, Die Wildkatzen der alten Welt. Eine Übersicht über die Untergattung Felis (Leipzig 1953).

1957: Th. Haltenorth, Die Wildkatze. Neue Brehm Bücherei 189 (Wittenberg-Lutherstadt 1957).

Hamilton 1896: E. Hamilton, The wild cat of Europe (*Felis catus*) (London 1896).

Hansen 2014: R. Hansen, Jagdzenen aus der Eifel. NaturFoto 2014/4, 26-33.

Hansen et al. 2001: J. E. Hansen / R. Ruedy / M. Sato / M. Imhoff / W. Lawrence / D. Easterling / T. Peterson / T. Karl, A closer look at United States and global surface temperature change. Journal of Geophysical Research 106/D20, 2001, 23947-23963.

Hartmann et al. 2013: S. A. Hartmann / K. Steyer / R. H. Kraus / G. Segelbacher / C. Nowak, Potential barriers to gene flow in the endangered European wildcat (*Felis silvestris*). Conservation Genetics 14/2, 2013, 413-426.

Heidemann 1973: G. Heidemann, Weitere Untersuchungen zur Nahrungsökologie »wildernder« Hauskatzen (*Felis silvestris* f. *catus* Linné, 1758). Zeitschrift für Säugetierkunde 38, 1973, 216-224.

Heinrich 1992: U. Heinrich, Erkenntnisse zum Verhalten, zur Aktivität und zur Lebensraumnutzung der Europäischen Wildkatze *Felis silvestris silvestris* Schreber, 1777 [Diss. Univ. Halle Wittenberg 1992].

Hemmer 1984: H. Hemmer, Die Carnivoren aus den Travertinen von Weimar. Quartärbiologie 5, 1984, 409-425.

1993: H. Hemmer, *Felis silvestris* Schreber, 1977 – Wildkatze. In: M. Stubbe / F. Krapp (eds), Handbuch der Säugetiere Europas 5: Raubsäuger – Carnivora (Fissipedia); II: Mustelidae 2, Viverridae, Herpestidae, Felidae (Wiesbaden 1993) 1076-1118.

Heptner/Sludskij 1980: V. G. Heptner / A. A. Sludskij, Wildkatze, *Felis* (*Felis*) *silvestris* Schreber, 1777. In: V. G. Heptner / N. P. Naumov (eds), Die Säugetiere der Sowjetunion. III: Raubtiere (Feloidea) (Jena 1980) 318-393.

Herrmann 1991: M. Herrmann, Säugetiere im Saarland – Verbreitung, Gefährdung, Schutz (Ottweiler 1991).

Hertwig et al. 2009: S. T. Hertwig / M. Schweizer / S. Stepanow / A. Jungnickel / U. Böhle / M. S. Fischer, Regionally high rates of hybridization and introgression in German wildcat populations (*Felis silvestris*, Carnivora, Felidae). Journal of Zoological Systematic and Evolutionary Research 47, 2009, 283-297.

von Hoberg 1687: W. H. von Hoberg, Georgica curiosa acuta. Anderer Theil, Elftes Buch (Nürnberg 1687).

Hossfeld/Reif/Reith 1993: E. Hossfeld / U. Reif / U. Reith, The Wildcat in the Taunus Mountains. Results of Preliminary Investigations and a Draft of a Research and Protection Project. Council of Europe Press, Environmental Encounters 16, 1993, 46-51.

Hötzel et al. 2007: M. Hötzel / N. Klar / S. Schröder / C. Steffen / C. Thiel, Die Wildkatze in der Eifel – Habitate, Ressourcen, Streifgebiete. Ökologie Säugetiere 5 (Bielefeld 2007).

Hu et al. 2014: Y. Hu / S. Hu / W. Wang / X. Wu / F. B. Marshall / X. Chen / C. Wang, Earliest evidence for commensal processes of cat domestication. Proceedings of the National Academy of Sciences of the United States of America 111/1, 2014, 116-120. DOI: 10.1073/pnas.1311439110.

Hubbard et al. 1992: A. L. Hubbard / S. McOrist / T. W. Jones / R. Boid / R. Scott / N. Easterbee, Is survival of European wildcats *Felis silvestris* in Britain threatened by interbreeding with domestic cats. Biological Conservation 61/3, 1992, 203-208.

Hucht-Ciorga 2011: I. Hucht-Ciorga, Wildkatzen: Heimliche Rückkehrer in unsere Wälder. Rheinisch-Westfälischer Jäger 7, 2011, 7-9.

Hupe/Simon 2007: K. Hupe / O. Simon, Die Lockstockmethode – eine nicht invasive Methode zum Nachweis der Europäischen Wildkatze (*Felis silvestris silvestris*). Informationsdienst Naturschutz Niedersachsen 27, 2007, 66-69.

Hupe/Pott-Dörfer/Götz 2004: K. Hupe / B. Pott-Dörfer / M. Götz, Nutzung autobahnnaher Habitate im Bereich der BAB 7 nördlich von Seesen durch die europäische Wildkatze (*Felis silvestris silvestris*) unter dem Aspekt der Lebensraumzerschneidung. Informationsdienst Naturschutz Niedersachsen 24, 2004, 266-278.

Jerosch/Götz 2011: S. Jerosch / M. Götz, Ist die offene Kulturlandschaft ein Wildkatzenlebensraum? Erste Ergebnisse einer Telemetriestudie in einem Verbundlebensraum. Beiträge zur Jagd und Wildforschung 36, 2011, 369-376.

Johnson/O'Brien 1997: W. E. Johnson / S. J. O'Brien, Phylogenetic Reconstruction of the Felidae Using 16S rRNA and NADH-5 Mitochondrial Genes. Journal Molecular Evolution 44/1, 1997, 98-116.

Johnson et al. 2006: W. A. Johnson / E. Eizirik / J. Pecon-Slattery / W. J. Murphy / A. Antunes / E. Teeling / S. J. O'Brien, The Late Miocene Radiation of Modern Felidae: A Genetic Assessment. Science 311, 2006, 73-77.

Jost 1978: H. Jost, Über die Verbreitung der Wildkatze (*Felis silvestris* Schreber) in Osthessen und Nachbargebieten. Beiträge zur Naturkunde in Osthessen 13/14, 1978, 81-99.

Kitchener et al. 2005: A. C. Kitchener / N. Yamaguchi / J. M. Ward / D. W. Macdonald, A diagnosis for the Scottish wildcat (*Felis silvestris*): a tool or conservation action for a critically-endangered felid. Animal Conservation 8, 2005, 223-237.

Knapp/Kluth/Herrmann 2002: J. Knapp / G. Kluth / M. Herrmann, Wildkatzen in Rheinland-Pfalz. Naturschutz bei uns 4, 2002, 1-24.

Kock/Altmann 1999: D. Kock / J. Altmann, Die Wildkatze (*Felis silvestris* Schreber 1777) im Taunus. Jahrbücher des Nassauischen Vereins für Naturkunde 120, 1999, 5-21.

Kratochvíl/Kratochvíl 1970: J. K. Kratochvíl / Z. Kratochvíl, Die Unterscheidung von Individuen der Population *Felis s. silvestris* aus den Westkarpaten von *Felis* s. f. *catus*. Zoologické Listy 19, 1970, 293-302.

Kratochvíl 1973: Z. Kratochvíl, Schädelkriterien der Wild- und Hauskatze (*Felis silvestris silvestris* Schreber 1777 und *Felis* s. f. *catus* L. 1758). Acta Scientiarum Naturalium Academiae Scientiarum Bohemoslovacae 7, 1973, 1-50.

Krug et al. 2012: A. Krug / H. Brede / F. Scharringhausen / U. Siebert / G. Gräber / G. Sodeikat, Erste Ergebnisse einer Telemetriestudie über Wildkatzen im Deister (Region Hannover) – ein Lebensraum für Wildkatzen auf dem Weg in den Norden. Säugetierkundliche Informationen 8/45, 2012, 377-385.

Krüger/Griebsch/Fischer 2015: M. Krüger / S. Griebsch / M. S. Fischer, Reichen Körpermaße für die Bestimmung von Wildkatzen und Hauskatzen sowie deren Hybriden aus? Landschaftspflege und Naturschutz in Thüringen 52, 2015, 23-28.

Krüger et al. 2009: M. Krüger / S. T. Hertwig / G. Jetschke / M. S. Fischer, Evaluation of anatomical characters and the question of hybridization with domestic cats in the wildcat population of Thuringia, Germany. Journal of Zoological Systematics and Evolutionary Research 47, 2009, 268-282.

Lecis et al. 2006: R. Lecis / M. Pierpaoli / Z. S. Biró / L. Szemethy / B. Ragni / F. Vercillo / E. Randi, Bayesian analyses of admixture in wild and domestic cats (*Felis silvestris*) using lined microsatellite loci. Molecular Ecology 15, 2006, 119-131.

de Leuw 1976: A. de Leuw, Die Wildkatze. Deutscher Jagdschutz-Verband, Niederwildausschuß: Merkblatt 16 (Bonn ³1976).

Lozano et al. 2003: J. Lozano / E. Virgós / A. F. Malo / D. L. Huertas / J. G. Casanovas, Importance of scrub-pastureland mosaics for wild-living cats occurrence in a Mediterranean area: implications for the conservation of the wildcat (*Felis silvestris*). Biodiversity and Conservation 12, 2003, 921-935.

Meinig 2002: H. Meinig, Erste Ergebnisse von Mageninhaltesanalysen bei 8 Wildkatzen (*Felis silvestris*) aus West-Deutschland mit Hinweisen zur Artbestimmung. Säugetierkundliche Informationen 26, 2002, 211-217.

Ministerium für Umwelt und Naturschutz, Landwirtschaft und Verbraucherschutz des Landes Nordrhein-Westfalen 2007: Geschützte Arten in Nordrhein-Westfalen. Vorkommen, Erhaltungszustand, Gefährdungen, Maßnahmen. Wildkatze (Düsseldorf 2007) 83-84.

Mitchell-Jones et al. 1999: A. J. Mitchell-Jones / G. Amori / W. Bogdanowicz / B. Krystufek / P. J. Reijnders / F. Spitzenberger / M. Stubbe / J. B. M. Thissen / J. Zima, The Atlas of European mammals (London 1999).

Moleón/Gil-Sánchez 2003: M. Moléon / J. M. Gil-Sánchez, Food habits of the wild cat (*Felis silvestris*) in a peculiar habitat: the Mediterranean high mountain. Journal of Zoology 260, 2003, 17-22.

Mölich/Klaus/Nöllert 2003: T. Mölich / S. Klaus unter Mitarbeit von A. Nöllert, Die Wildkatze in Thüringen. Landschaftspflege und Naturschutz in Thüringen 40, 2003, 109-135.

Müller 2005: F. Müller, Zur Diagnostik von Wild- und Hauskatzen (*Felis silvestris* und *F. catus*, Felidae) nach morphologischen und anatomischen Merkmalen. Beiträge zur Naturkunde in Osthessen 41, 2005, 9-18.

2011a: F. Müller, Zur individuellen Variabilität von Körpermerkmalen, insbesondere des Fell-Zeichnungsmusters bei adulten Wildkatzen (*Felis s. silvestris*, Felidae) aus Mitteluropa. Beiträge zur Jagd und Wildforschung 36, 2011, 351-358.

2011b: F. Müller, Körpermerkmale als Unterscheidungskriterien zwischen wildfarbenen Hauskatzen (*Felis s. catus*) und Wildkatzen (*Felis silvestris silvestris*, Felidae) aus Mitteleuropa. Beiträge zur Jagd und Wildforschung 36, 2011, 359-368.

NABU Bundesverband 2007: Der NABU-Bundeswildwegeplan (Bonn 2007).

Nussberger 2012: B. Nussberger, Introgression zu erkennen ist wichtig für den Wildkatzenschutz. Säugetierkundliche Informationen 8/45, 2012, 319-327.

Nussberger/Wandeler/Camenisch 2014: B. Nussberger / P. Wandeler / G. Camenisch, A SNP chip to detect introgression in wildcats allows accurate genotyping of single hairs. European Journal of Wildlife Research 60, 2014, 405-410.

Nussberger et al. 2014: B. Nussberger / P. Wandeler / D. Weber / L. F. Keller, Monitoring introgression in European wildcats in the Swiss Jura. Conservation Genetics 15, 2014, 1219-1230.

Oliveira et al. 2008: R. Oliveira / R. Godinho / E. Randi / N. Ferrand / P. C. Alves, Molecular analysis of hybridisation between wild and domestic cats (*Felis silvestris*) in Portugal: implications for conservation. Conservation Genetics 9, 2008, 1-11.

Pflüger 1987: H. Pflüger, Die Wildkatze in Hessen. Merkheft zum Schutz der Wildkatze. Schriftenreihe für Natur und Umweltschutz 1 (Frankfurt [Main] 1987).

Piechocki 1987: R. Piechocki, Historischer Nachweis einer Wildkatze (*Felis silvestris*) bei Halle (Assle). Hercynia N. F. 24, 1987, 464-465.

1989: R. Piechocki, Wildkatze *Felis silvestris* Schreber. In: H. Stubbe (ed.), Buch der Hege. 1: Haarwild (Berlin 1989) 342-372.

1990: R. Piechocki, Die Wildkatze *Felis silvestris*. Neue Brehm Bücherei 189 (Wittenberg-Lutherstadt 1990).

Piechocki/Möller 1983: R. Piechocki / H. Möller, Über das Ansprechen, den Schutz und die Lebensweise der Wildkatze I, II, Schluß. Unsere Jagd 33, 1983, 14-15. 52-53. 82-83.

1991: R. Piechocki / H. Möller, Zur Biologie und Verbreitung der Wildkatzen im Harz und Thüringer Wald. Wiesenfelder Reihe 8, 1991, 52-59.

Piechocki/Stiefel 1988: R. Piechocki / A. Stiefel, Über die Altersstruktur der Verluste der Wildkatze (*Felis silvestris* Schreber 1777). Hercynia N. F. 25, 1988, 235-258.

Plos 2015: J. M. Plos, Die Mensch-Katzen-Beziehung im mittelalterlichen christlichen Abendland unter besonderer Berücksichtigung ambivalenter Aspekte und deren Ursachen [Diplomarbeit Univ. Graz]. http://unipub.uni-graz.at/obvugrhs/download/pdf/454981?originalFilename=true (12.5.2020).

Pott-Dörfer/Dörfer 2007: B. Pott-Dörfer / K. Dörfer, Zur Ausbreitungstendenz der Wildkatze *Felis silvestris silvestris* in Niedersachsen. – Ist die niedersächsische Wildkatzenpulation gesichert? Informationsdienst Naturschutz Niedersachsen 27, 2007, 56-62.

Ragni/Possenti 1996: B. Ragni / M. Possenti, Variability of coat-colour and markings system in *Felis silvestris*. Italian Journal of Zoology 63, 1996, 285-292.

Randi/Ragni 1991: E. Randi / B. Ragni, Genetic variability and biochemical systematics of domestic and wild cat populations (*Felis silvetris*: Felidae). Journal of Mammalogy 72/1, 1991, 79-88.

Reig/Daniels/Macdonald 2001: S. Reig / M. J. Daniels / D. W. Macdonald, Craniometric differentiation within wild-living cats in Scotland using 3D morphometrics. Journal of Zoology 253, 2001, 121-132.

Raimer 1994: F. Raimer, Die aktuelle Situation der Wildkatze in Deutschland. Wiesenfelder Reihe 13, 1994, 15-39.

2001: F. Raimer, Heimlichkeit in weiten Wäldern – Der Schutz der Wildkatze und ihrer Lebensräume. In: H. Grabe / G. Worel (eds), Die Wildkatze zurück auf leisen Pfoten (Amberg 2001) 71-89.

2006: F. Raimer, Die Wildkatzenpopulationen in Hessen und Niedersachsen seit dem 18. Jahrhundert – Verfolgung, Bedrohung, Schutz und Wiederausbreitung. In: Naturschutz-Akademie Hessen, Bund für Umwelt und Naturschutz Deutschland, Institut für Tierökologie und Naturbildung (ed.), Kleine Katzen – große Räume: ein Rettungsnetz für die Wildkatze. Naturschutz-Akademie Hessen (Wetzlar): Akademie-Berichte 5 (Wetzlar 2006) 69-78.

Röben 1974: P. Röben, Die Verbreitung der Wildkatze, *Felis sylvestris* Schreber, 1777, in der Bundesrepublik Deutschland. Säugetierkundliche Mitteilungen 22, 1974, 244-250.

Schauenberg 1969: P. Schauenberg, L'identification du chat forestier d'Europe, *Felis s. silvestris* Schreber 1777, par une méthode ostéométrique. Revue Suisse de Zoologie 76, 1969, 433-441.

1970: P. Schauenberg, Le chat forestier d'Europe *Felis s. silvestris* Schreber 1777 en Suisse. Revue Suisse de Zoologie 77, 1970, 127-160.

1977: P. Schauenberg, Longueur de l'intestin du chat forestier *Felis silvestris* Schreber. Mammalia 41, 1977, 357-360.

Scott/Easterbee/Jefferies 1993: R. Scott / E. N. Easterbee / D. Jefferies, A Radio-Tracking Study of Wildcats in Western Scotland. Council of Europe Press, Environmental Encounters 16, 1993, 94-97.

Simon/Raimer 2007: O. Simon / F. Raimer, Wanderkorridore von Wildkatze und Rothirsch und ihre Relevanz für künftige infrastrukturelle Planungen in der Harzregion. Informationsdienst Naturschutz Niedersachsen 27, 2007, 27-37.

Sládek 1973a: J. Sládek, Jahreszeitliche und jahresbedingte Veränderungen der Nahrung der Wildkatze (*Felis silvestris*, Schreber 1777) in den Westkarpaten. Zoologické Listy 22, 1973, 127-144.

1973b: J. Sládek, Beitrag zur Kenntnis der Ernährung der Wildkatze (*Felis silvestris* Schreber 1777) in den Westkarpaten. Lynx 14, 1973, 38-53.

Sládek/Mošanský/Palášthy 1971: J. Sládek / A. Mošanský / J. Palášthy, Die Variabilität der Schädelkapazität bei der Westkarpaten-Population der Wildkatze, *Felis silvestris* Schreber, 1777. Zoologické Listy 20, 1971, 153-160.

1972: J. Sládek / A. Mošanský / J. Palášthy, Variabilität der linearen kraniologischen Merkmale bei der westkarpatischen Population der Wildkatze, *Felis silvestris* Schreber, 1777. Zoologické Listy 21, 1972, 23-37.

Sodeikat/Kögelsperger 2012: G. Sodeikat / P. Kögelsperger, Monitoring der Wildkatzenpopulation (*Felis silvestris silvestris* Schreber, 1777) im Elm. Säugetierkundliche Informationen 8/45, 2012, 373-376.

Sommer/Benecke 2006: R. S. Sommer / N. Benecke, Late Pleistocene and Holocene development of the felid fauna (Felidae) of Europe: a review. Journal of Zoology 269, 2006, 7-19.

Splitter 1978: H. Splitter, Untersuchungen zur Nahrungsbiologie streunender Hauskatzen (*Felis sylvestris* f. *catus* L.). Zeitschrift für Jagdwissenschaft 24, 1978, 35-44.

Stahl/Artois 1995: P. Stahl / M. Artois, Status and conservation of the wildcat (*Felis silvestris*) in Europe and around the Mediterranean rim. Council of Europe Nature and Environment Series 69, 1995, 1-77.

Stefen 2012a: C. Stefen, Craniometric study of wild cats (*Felis silvestris,* Carnivora: Felidae) in Germany – are there differences between populations or in time. Acta Scientiarum Naturalium Academiae Scientiarum Bohemoslovacae 76, 2012, 127-162.

2012b: C. Stefen, Gab es Veränderungen in der Nahrung der Europäischen Wildkatze (*Felis silvestris* Schreber, 1777) in den letzten Jahrhunderten? Säugetierkundliche Informationen 45, 2012, 347-372.

2015: C. Stefen, Does the European wildcat (*Felis silvestris* Schreber, 1777) show a change in weight and size with global warming? Folia Zoologica 64, 2015, 65-78.

Stefen/Görner 2009: C. Stefen / M. Görner, Wildkatze in Deutschland und Mitteleuropa – zum Stand der Forschung und Konsequenzen für den Schutz. Säugetierkundliche Informationen 38, 2009, 1-216.

Stefen/Heidecke 2011: C. Stefen / D. Heidecke, Kraniometrische Variabilität der Wildkatze (*Felis silvestris* Schreber, 1777) im Harzgebiet. Hercynia N. S. 44/2, 2011, 253-285.

Stefen/Görner/Schmidt 2009: C. Stefen / M. Görner / T. Schmidt, Europäische Wildkatze *Felis silvestris*. In: M. Görner (ed.), Atlas der Säugetiere Thüringens. Biologie, Lebensräume, Verbreitung, Gefährdung, Schutz (Jena, Erfurt 2009) 224-227.

Steyer et al. 2013: K. Steyer / O. Simon / R. H. Kraus / P. Haase / C. Nowak, Hair trapping with valerian-treated lure sticks as a tool for genetic wildcat monitoring in low-density habitats. European Journal of Wildlife Research 59, 2013, 39-46.

Streif et al. 2012: S. Streif / S. Kraft / S. Veith / A. Kohnen / R. Suchant, Die Wildkatze (*Felis silvestris*) in Baden-Württemberg – Statusbericht zu Monitoring und Forschung. Säugetierkundliche Informationen 8/45, 2012, 411-416.

Stubbe/Stubbe 1994: M. Stubbe / A. Stubbe, Säugetierarten und deren feldökologische Erforschung im östlichen Deutschland. Tiere im Konflikt 1994/3, 1994, 3-52.

Vigne et al. 2004: J. D. Vigne / J. Guilaine / K. Debue / L. Haye / P. Gérard, Early taming of the cat in Cyprus. Science 304/5668, 2004, 259-259.

Vogt 1985: D. Vogt, Aktuelle Verbreitung und Lebensstätten der Wildkatze (*Felis silvestris silvestris* Schreber 1777) in den linksrheinischen Landesteilen von Rheinland-Pfalz und Beiträge zu ihrer Biologie. Beiträge zur Landespflege Rheinland Pfalz 10, 1985, 130-165.

Weigel 1972: I. Weigel, Kleinkatzen und Nebelparder. In: R. Altevogt / R. Angermann / H. Dathe / B. Grzimek / K. Herter / D. Müller-Using / U. Rahm / E. Thenius, Grzimeks Tierleben. 12: Säugetiere III (Zürich 1972) 287-333.

Witticke 2015: H. Witticke, Thüringens Wälder – ihr Werden, ihr Wachsen und ihre Naturschutzgeschichte. In: M. Görner (ed.), Thüringen: Wald und Wild, Gewässer und Fische, Landschaften und Arten (Gera 2015) 8-101.

Wittmer 2001: H. U. Wittmer, Home range size, movements, and habitat utilization of three male European wildcats (*Felis silvestris* Schreber, 1777) in Saarland and Rheinland-Pfalz (Germany). Mammalian Biology 66, 2001, 365-370.

Yamaguchi et al. 2004: N. Yamaguchi / C. A. Driscoll / A. C. Kitchener / J. M. Ward / D. W. MacDonald, Craniological differentiation between European wildcats (*Felis silvestris silvestris*), African wildcats (*F. s. lybica*) and Asian wildcats (*F. s. ornata*): implications for their evolution and conservation. Biological Journal of the Linnean Society 83, 2004, 47-63.

SUMMARY / ZUSAMMENFASSUNG

The European Wildcat *Felis silvestris* – A Brief Review of Status and Research in Germany

The European wildcat is a Palaearctic species which probably evolved in the cooler periods of the Late Pleistocene. It is morphologically very similar and phylogenetically close to but not the direct ancestor of the domestic cat. Like other carnivores it has been strongly hunted prosecuted particularly in the 17th to 19th century and nearly extirpated in the early 20th century. In Germany it was first protected in 1934 and has been under special protection under the European Habitats Directive ever since then. This interesting history is reflected in a substantial amount of literature in zoology, hunting and ecology. This article presents a brief review on the most important aspects of their morphology, differentiation from domestic cats, ecology including diet, reproduction and habitat, as well as endangerment, population status and protection.

Die Europäische Wildkatze *Felis silvestris* – ein kurzer Rückblick auf Stand und Forschung in Deutschland

Die Europäische Wildkatze ist eine paläarktische Art, die sich wahrscheinlich in den kühleren Perioden des Spätpleistozäns entwickelt hat. Sie ist morphologisch sehr ähnlich und phylogenetisch nahe, aber nicht der direkte Vorfahre der Europäischen Wildkatze. Wie andere Fleischfresser wurde sie im 17.-19. Jahrhundert stark gejagt und im frühen 20. Jahrhundert fast ausgerottet. In Deutschland wurde sie erstmals 1934 geschützt und steht unter dem besonderen Schutz der Europäischen Habitatrichtlinie. Diese interessante Geschichte spiegelt sich in einer beträchtlichen Menge an Literatur in Zoologie, Jagd und Ökologie wider. Dieser Artikel gibt einen kurzen Überblick über die wichtigsten Aspekte ihrer Morphologie, Unterscheidung von Hauskatzen, Ökologie einschließlich Ernährung, Fortpflanzung und Lebensraum sowie Gefährdung, Populationsstatus und Schutz.

FRITZ JANTSCHKE

TIGER AND MAN – THE DIFFICULT CO-EXISTENCE OF TWO TOP PREDATORS IN INDIA

»It is the most terrible of all cats, a predator against which so far even man is still powerless.« An insistent description of the tiger by Alfred Brehm in his »Thierleben« from 1876.
»No predatory mammal can combine so much horror with truly seductive beauty. Man has exaggerated its bloodlust or at least portrayed it with very utilising colours. However, this is not surprising, because for those who could describe it, it is the sum total of all cruelty. Even today a frightening number of tigers inhabit India, and thousands of people have to be mobilized even to temporarily free a region which would otherwise fall prey to desolation, from this worst of all plagues.«
What Alfred Brehm wrote more than a century ago is no longer true. The tiger is – apart from a single region – no menace anymore. By now its numbers are so greatly reduced that it is in danger of becoming extinct. Additionally, our attitude towards this beautiful big cat (**fig. 1**) has changed, as it has become one of the most popular animal figures. This is obvious not only by the use of tigers in advertisements, but also in the promotion of tourist destinations like »Tigerland«. To look into a tiger's eyes is way better than a photo safari for lions or elephants in Africa.

Fig. 1 »The most terrible of all cats« – that's how Alfred Brehm described the tiger over 100 years ago. Today we see the big cat in a completely different light. – (All photos F. Jantschke).

In: Nicholas J. Conard · Henning Hassmann · Kurt Felix Hillgruber · Jordi Serangeli · Thomas Terberger (eds), The *Homotherium* Finds from Schöningen 13II-4. Man and Big Cats of the Ice Age. Forschungen zur Urgeschichte aus dem Tagebau von Schöningen 4 (Mainz 2022). DOI: https://doi.org/10.11588/propylaeum.1006.c13529

Fig. 2 To look a tiger in the eye is the dream of many tourists visiting India's tiger reserves today.

We do not want to deal with the tiger primarily from the perspective of tourism (**fig. 2**). Two questions are more important: Does the big cat have a chance of surviving in a densely populated India? And how can people coexist with a predator that is not only dangerous but sometimes known as a »man-eater« who specifically even hunts them?

The tiger is undoubtedly one of the most attractive and beautiful mammals. With its strikingly striped fur, a size of approximately 2.5 m without tail and a weight of nearly 300 kg, this big cat blends remarkably well into its surroundings, whether it is dry grass, leaves, bamboo jungle or light *śāl*-forest (*Shorea robusta*). In nature, tigers are much harder to detect than the dun-coloured lions. But – and this is a positive development – tigers have become much more visible. At least in the Indian national parks it is no longer necessary for tigers to hide from people. They are encountered in broad daylight in the middle of a street or taking a nap on the lake shore without worrying about vehicles or photographers.

Only 40-50 years ago this was totally different. Tigers were hunted continuously, often as a feudal amusement. The rich and powerful Maharajas made uninhibited use of their privileges. This included the tiger hunt, which was operated under an immense output of manpower and material (**fig. 3**). Some Maharajas

Fig. 3 Representation of an old hunting scene (Maharaja hunt). The Maharajas and their guests undertook the tiger hunt with great enthusiasm, often by riding an elephant´s back.

were known to have hunted more than 1000 tigers. In the British colonial period, senior officials also took to the hunting of tigers too. Towards the end of colonial rile, the British viceroy Lord Linlithgow shot, in one single hunting season in 1938-1939, 120 tigers. With India's independence, the situation did not improve, it got even worse. The number of hunters increased dramatically. Wealthy Indians, Europeans, Americans, all wanted to kill a tiger, to have a photo of it over, or a tiger fur in front of the fireplace.

The majority of Indians, without the means for sportsmanlike hunting, could at least afford poison or a trap to kill a tiger. Either to slay a »rustler« or to enrich themselves by selling fur or bones on the black market.

Fig. 4 The first director of the ambitious and successful Indian »Project Tiger« was the former zoo director Kailash Sankhala.

The demise of the tigers came also indirectly. The dramatic growth of the human population – with almost 1.3 billion people India is by now more densely populated than Germany – invaded nearly all the habitats of the striped cats. They are now mostly used for agricultural purposes, from fields via pastures to plantations. Large carnivores have no room anymore. The forests in India are often totally exploited. Intensive livestock farming leads to overgrazing and bare soils with erosion washing out the roots of the trees. Branches are constantly cut off as fodder for livestock. No suitable habitat remains for deer and other prey species of tigers.

It is no surprise that with the decline of habitats and resources the number of tigers in India decreased drastically. Allegedly, at the beginning of the last century, around 100 000 of them lived in India. Since 1940, around the time of India's independence, their number was still thought to be approx 30 000. In the mid-1960s, it had dropped to only a few thousand. The first appeals to protect the tiger were raised. In 1969 the dramatic situation was clearly addressed at an international conservation conference in Delhi. The Indian government under Indira Gandhi reacted promptly and banned the hunting of tigers and the export of tiger skins the following year.

In 1973 »Project Tiger« was started under the leadership of Kailash Sankhala (**fig. 4**). He had worked in the state forest services and became the zoo director of Delhi in 1965. His first task was to carry out a nationwide tiger census. The common method for this at that time was the counting of tiger tracks, as these vary for each animal (**fig. 5**). The result was even more devastating than expected: In the whole of India only 1827 tigers were counted. Within three decades, the country had lost at least 25 000 of the big cats.

For the protection of their tigers, India has spent – up to the present day – an immense fortune: according to experts about 4 million dollars per year. In addition the annual loss of revenue due to the renunciation of the timber industry and other uses were summed up to a total of around 14 million dollars. In 1973 the first nine tiger reserves were established, encompassing approximately 10 000 km² in total. Ten years later the number increased to 15 parks with an area of 25 000 km². Within this period, the number of tigers in the reserves increased threefold, and the total population in India doubled.

Many of the first tiger reserves were former hunting grounds of the Maharajas, for example Ranthambhore, Sariska and Bandhavgarh, to name just three of the most important ones. These rulers, even when

Fig. 5 Tiger's footprint. Even today, the common form of the determination of a tiger population is the counting of the different tiger tracks.

compared to feudal conditions in Europe, had a staggering amount of power and wealth. The state of their indulgences can be estimated by the magnificence of the hunting lodge Sariska, now a tourist hotel, which the former owner called »my little shooting box« (**fig. 6**). The Swedish nature photographer Bengt Berg created a literary monument to this hunting lodge and the Maharaja in his book »Tiger and Man«. It was written in 1934 when these rulers were still in power. A short excerpt from this work: »Whoever is able to provide his esteemed guests with a tiger as a present has a trump in his hand. Money can be used to bribe social inferiors. With a tiger one can show his generosity towards superiors. Therefore, a well-kept tiger hunting preserve is a capital, more valuable to his princely owner than the tax documents of his state. As a good host, he hardly dares to shoot a tiger himself anymore, because a Maharaja without a tiger is lacking something that cannot be replaced by gold.«

The Kanha National Park, one of the most famous tiger reserves, was at least partly the hunting ground of the British viceroys. Right at the beginning of the »Project Tiger«, the reserve in the Central Indian Highlands was enlarged from 318 to 448 km², two years later it doubled in size. There was also a buffer zone

Fig. 6 Sariska Palace. »My little shooting box« is what the Maharaja of Alwar called the magnificent castle in his hunting ground Sariska, today a tourist hotel.

Fig. 7 The Kanha National Park, a tiger reserve from the very beginning, is mainly shaped by *śāl*-forest.

of 1000 km² incorporated. Much of the area is covered by evergreen *śāl*-forest (**fig. 7**), but there are also regions with deciduous monsoon forest. Another element of the vegetation are bamboo thickets, situated especially on hillsides and riverbanks, perfect hiding places for the tiger (**fig. 8**).

The most common inhabitants of this light, open forest are axis deer (*Axis axis*). One of the most prestigious and rarest animals in the reserve is the impressive gaur (*Bos gaurus*), the largest living bovine species. It lives predominantly in forests and feeds on leaves and herbs. Another rare species of the parklands is the Central Indian Barasingha deer (*Cervus duvauceli branderi*). At the beginning of the »Project Tiger«, only about 60 of them still survived in Kanha. After a study by the Swiss biologist Claude Martin, their living conditions were improved and their stock increased tenfold in just under 20 years. Key measures were the enlargement of the national park and the relocation of people. In the main habitats of the Barasinghas former rice terraces are still visible everywhere. Among the many activities of the national parks is the preservation of water reservoirs formerly established for the livestock. In the dry season, fallen leaves and twigs are constantly being swept from the paths and burned, thus creating firebreaks and preventing forest fires from one side to the other.

Fig. 8 By sweeping the paths and burning the dry leaves, effective firebreaks are created in the Kanha Tiger Reserve.

Fig. 9 Tigers in front of riding elephants with tourists. The tigers are hardly impressed, as they are visited for one or two hours by riding elephants with tourists in some reserves.

Fig. 10 Tigers with jeeps. Tigers also casually stroll past jeeps full of foreign and local tourists.

The infrastructure for visiting the Kanha Tiger Reserve, mainly by small Indian jeeps, is well developed. The best way to get to know the habitat off the beaten track is on the backs of elephants. Unfortunately, there are not enough of the proboscideans for all the tourists. Therefore, they are primarily used to take visitors directly to the tigers. Each morning rangers search at suitable locations for tigers which are accustomed to humans and elephants. After positioning elephants there, they are then used to shuttle tourists to and from their jeeps in short rides to the big cats (**figs 9-10**).

On Sundays, between 9 and 11 in the morning Kanha is visited by many locals and a few hundred people will gather, waiting to be carried by three to five elephants to the chosen tiger sites. This way they are normally getting better photos than from the jeep because elephants are allowed to leave the tracks and thus can bring the tourists very close to the tigers. These are accustomed to this procedure and show no sign of disturbance. The tiger population in the park is relatively constant and estimated at around 100 animals. The great interest of foreign and native tourists contributes significantly to their conservation.

Essential research on the territorial behavior of tigers was done at the Kanha preservation site leading to new unique discoveries:

1. All tigers favor areas with a high prey density. The strongest animals succeed. Weaker individuals, primarily the young and old, have to move to poorer areas. In the core areas each animal has its own, strictly defined territory.
2. Areas of high-ranking males may include one or more females. Their territories may overlap slightly.
3. The territorial males have exclusive mating rights with these females and grant their protection to their offspring.
4. Areas with a good supply of prey are also the main birthplaces.
5. Animals with deteriorating health live in more peripheral areas and their territories may not be clearly defined any more.
6. Old and young tigers have to emigrate to the poorest areas.
7. In large forest areas these emigrating tigers can provide a genetic exchange between populations. But many of them fall prey to poachers.

True wildlife areas are concentrated mainly in the mountain region, from the Himalayas to the Southern Indian Nilgiris. By contrast, in the lower elevations a »green revolution« led to a modest surplus of food production. Therefore, agriculture in less favorable areas has decreased, which has had positive effects on the regeneration of natural habitats.

The conservational goals have already been implemented to quite a large extent. In 1975 there was a total of five national parks and 126 other reserves covering an area of 24000 km². Eleven years later the numbers have increased to 54 national parks and 248 reserves, covering a total of 100000 km². Today, India has more than 100 national parks, and the number of tiger reserves has grown from the original nine to 49. Unfortunately, the population of the tiger has not increased to the same extent. On the contrary, in 2006 an accurate survey concluded that only 1418 tigers were left in India, less than at the beginning of the »Project Tiger«. In various regions, even in national parks, poaching has been intense. No wonder, since the bones of the tiger are virtually worth their weight in gold in Traditional Chinese Medicine. The fact that the number of tigers has been corrected two years later to 1706, puts the reliability of official data in doubt. One had forgotten to include one of the most important areas for the tiger population, the Sundarbans on the border between India and Bangladesh. The current estimate is about 2400 living tigers for the whole of India.

One of the most famous and popular tiger reserves is Ranthambhore, also a former Maharaja hunting reserve. The name derives from the ruin of a fortress that overlooks the entrance to the reserve and can be seen from afar. In many parts of the park there are remains of ancient settlements, for example, small Hindu

temples. This together with lakes, meadows and dry bushy forests creates much of the charm of Ranthambhore. Without the lakes, which are mostly artificial, the animal wealth of axis (*Axis axis*) and sambar deer (*Cervus unicolor*) and many other species would be inconceivable. Only a short distance away from the water sources, the landscape is, outside the monsoon season, mostly dry.
The tiger reserve of Ranthambhore was founded and led for two decades by Fateh Singh Rathore. Under his administration the small 400 km² reserve became an excellent tiger habitat and he himself a renowned tiger expert. We owe much of our knowledge about the social life of tigers to him. Before Rathore, we were unaware that tiger offspring will remain for up to two years with the mother. Male tigers are no threat for the young ones. Many even visit their offspring and the mothers regularly. Tigers are much more social animals than was previously thought possible.
For the prey of the tiger Fateh Singh came up with the following order: The primary quarry are not the very common axis deer (*Axis axis*), but wild boars (*Sus scrofa cristatus*). They are often so busy searching the ground for food that they easily fall prey to the cats. According to Fateh Singh, the careless sambar deer (*Cervus unicolor*) are the second most likely victims of the large cats. If they discover something suspicious, they do not flee immediately but come closer and take a curious, and often their last, look. In third place are the large, also relatively careless nilgai antelope (*Boselaphus tragocamelus*). Only fourth as tiger prey are the axis deer (*Axis axis*), because they are extremely vigilant and always have some sentinels positioned in their herds. But basically tigers prey on nearly every conveniently available animal, from the gray langur (*Semnopithecus entellus*) to the gaur (*Bos gaurus*), occasionally even young elephants and rhinos. And sometimes also people …
With the exception of humans, tigers have no real enemies, only competitors like the leopard (*Panthera pardus*). In tiger areas it is rarely seen during the day and seems to avoid the larger rival. A pack of the very social dholes (*Cuon alpinus*) can probably overwhelm a tiger, but they are also food competitors. Other predators like the golden jackals (*Canis aureus*) or bears sometimes feed on a tiger's prey and therefore compel these big felids to go hunting again.
Towards the end of his career, Fateh Singh was transferred to Sariska Tiger Reserve in order to rehabilitate it. Instead of the 40 tigers officially reported there he was only able to verify the existence of 17 or 18 animals. A skeptical view on the reported figures is mainly due to a marked weak point in Indian bureaucracy: The directors of the national parks are regularly transferred to other positions. In order to prove their qualification they simply use a slightly larger number of animals than before. Verifying their information is hardly possible. Meanwhile, the stocks are therefore determined almost entirely by photo traps.
At the commencement of his duties in Sariska, Fateh Singh immediately closed all the observation posts at the waterholes, from which poachers had hunted at least 26 tigers. While such measures are easy to perform, the resettlement of villagers from the national parks is much more difficult. Since they cannot be forced to leave persuasion is necessary, for example by the promise of a better infrastructure such as schools, wells and public transport just outside the park. In the reserves people cause significant problems. They monopolize the best areas including the waterholes and their cattle and goats are not only a competition for the wildlife, but the habitats are overgrazed considerably. The results are severely degraded environments around the villages.
It is obvious that establishing reserves and creating new protection laws is not sufficient. Realizing this, a longtime fellow of Fateh Singh, the social anthropologist Valmik Thapar, drew consequences. Instead of focusing solely on tiger research and preservation he devoted himself increasingly to the assistance of local people which also indirectly benefitted the tigers. For this purpose the Ranthambhore Foundation was established. With the financial help of the foundation a protective wall against livestock was erected around the reserve. The villagers, who carried out this construction work themselves, may harvest only grass for their

cattle in the park. Instead they receive trees from a nursery and a mobile ambulance is now in use, which is vital for health care in the neighboring villages of the national park.

The foundation promotes local craftsmanship, for example the production of shoes and clothing. In a breeding station, there are also bulls of water buffalos (*Bubalus arnee*) and humped cattle (*Bos primigenius indicus*) for improving the livestock. With that measure the number of cattle can be kept down.

In addition to the villagers, there are also native tribes who have no interest in »progress« and development. For example, the Gond people still live in the area of the Kanha National Park and sustain their lives by what nature has to offer, for example, collecting wild fruit in the reserve. They hardly disturb the natural balance.

Besides Kanha and Ranthambhore, a number of additional parks have become excellent habitats for the big cats – the Corbett National Park in the foot hills of the Himalayas and Bandhavgarh situated in the Central Indian Madhya Pradesh province. They are also popular with nature tourists. Kaziranga National Park in northeastern Assam province, famous for its population of Indian rhinos (*Rhinoceros unicornis*), also contains a healthy tiger population. Similarly well-equipped for the last Indian lions (*Panthera leo persica*) is the Gir Forest north of Mumbai (formerly Bombay). No doubt, India has as much to offer in terms of nature as in its renowned culture.

The distribution of tigers, which formerly ranged over large parts of Asia, has shrunk considerably. Three of the nine subspecies have become extinct, a fourth is only surviving in captivity. The Bali tiger vanished before World War II, the Javan tiger in the mid-1970s and in the last years also the Caspian tiger. The South Chinese tiger is probably extinct. Only around 350 animals of the Indochinese tiger are still alive, about as many live in Sumatra (Sumatra tiger). Of the Malayan tiger, a subspecies only recently set-up, about 500 individuals are estimated to be alive. Only doubtful and fluctuating information on the Amur or Siberian tiger are available: around 400 animals are left. Including the large population of perhaps 3000 animals which live in India and some neighboring countries like Nepal and Bangladesh, only 4500 Bengal tigers are left.

In three subspecies the habitat they require is quite accurately determined. Not surprisingly the Siberian tiger roams through the largest areas. This is not only because of their enormous size but also because in their region the food supply is rather limited. The home range of females is around 200-400, for males between 800 and 1000 km². An area of 1000 km² can support about five to six tigers. There is a marked contrast with the favorable conditions in India, where 14 tigers thrive in an area of around 100 km², more than 20 times as many as in Siberia. Four to six individuals are expected to live in an area of 100 km² in Nepal, whereas only four Sumatran tigers can share that space. The Indian subcontinent seems to offer favorable conditions for the tigers.

A crucial question is: How do people in India cope with the tigers, which prey not only on their livestock, but also on themselves? After all Wikipedia (an admittedly rather dubious source) mentions that since the 1800s up to 373 000 people have been killed by tigers. Numbers which can hardly be verified. Much more reliable is a contribution by the American tiger expert Charles McDougal »The man-eating tiger in geographical and historical perspective«, which was published in 1986 in the book »Tigers of the World«. According to him, there were problems with tigers mainly in South China (in a village up to 60 people were killed in a few weeks in 1922), Singapore (where in the 1940s 200-300 people died) and in Manchuria. In the latter, the officials' response was swift and dealt with extreme severeness. A lot of incidents occurred in what is today India. In 1822 about 500 people were killed by wild animals, almost exclusively by tigers, in the Bombay district alone. In the foothills of the Himalayas »man-eaters« appeared repeatedly and became quite well-publicized by the books of Jim Corbett (**fig. 11**). And in the central provinces of Madhya Pradesh and Maharashtra 200-300 people were killed by tigers annually during the mid-19th century. Particularly dangerous was (and still is) Bengal, which today comprises the Indian district of West Bengal and western Bangladesh, where around 500 people were killed every year.

Fig. 11 The famous tiger hunter Jim Corbett (whose bust stands at the entrance of the national park named after him) was a pioneer of tiger protection.

The death rate by tigers in the British colonial territory (today's India, Pakistan and Bangladesh) was quite high in the early last century: From 1902 to 1911 an average of 850 people died every year. In the 1920s, the death rate was even higher and the year 1922 holds the record with 1603 deaths. From 1928 to 1933, there were more than 7000 deaths, an average of nearly 1200 per year. After World War II, when India became independent, the number of tigers was significantly reduced, resulting in the near absence of fatal attacks on humans in most areas. But since then there have always been exceptions: 128 deaths were recorded in the years 1978-1984 in the Uttar Pradesh province. These killings ended after ten tigers were slain and two were captured.

Jim Corbett, who earned himself the reputation of a tiger hunter, was always called in to help when a tiger problem started in any region. His conclusion was that in general animals that were wounded or unable to hunt natural prey due to old age or sickness were the problematic ones. But on the other side there are also disabled tigers which do not hunt humans and healthy tigers which specialize for some reason in hunting people. In the late 19th century, no problems with tigers were known in the border regions of the Himalayas, where Jim Corbett was mainly active at that time. That changed when their food supply had deteriorated significantly, leading to attacks on domestic animals and humans.

A real special case is Sundarbans (**figs 12-13**), an area of about 16000 km² comprising of the deltas of the Ganges, Brahmaputra and Meghna at the border between India and Bangladesh. It is still a region prone with man-hunters. 100 or more people fall prey to the big cats every year. In 1971, a first scientific study of the man-hunters was led by the zoologist Hubert Hendrichs. Recent data on this topic, especially from Bangladesh, derives from the Southeast Asia experts Gertrud and Helmut Denzau. After Hendrichs only 1 % of the tiger population (the estimated number in both countries is up to 500 animals, the largest, only truly viable population worldwide) is specialized in humans as prey. About 30 % of the attacks are accidental and the vast majority of tigers avoid humans.

There are various theories for the attacks of tigers on humans.

- The lack of natural prey may lead to attacks on livestock and consequently on shepherds (especially children).
- The high salinity of the drinking water could make tigers particularly aggressive. Therefore, freshwater ponds have been established.
- Due to the constant flooding of their habitat the territorial markings are washed away regularly. This may trigger more dominant/aggressive behavior.
- As a result of devastating cyclones a large number of people (sometimes thousands) perish. Tigers could find dead bodies and become accustomed to humans as a source of food.

Fig. 12 Sundarbans. The mangrove area of Sundarbans in the delta of the mouth of three large rivers is in many places only navigable by boat. Even today tigers fall victim to fishermen in Sundarbans on the border to Bangladesh – and vice versa.

- The wet and slippery habitat makes it difficult to hunt. In comparison, man is easier to capture than natural prey.
- The people who go about their work foraging in the marshes are often bent down to catch fish or shrimp and gather wood and thus evoke the predatory instincts of tigers more.
- People do not live directly in the marshes and are often not accustomed to dealing with the big cats.
- In other areas, the tigers were more intensively hunted, reduced in number and became wary of humans.

Fig. 13 Tiger tracks. These tracks at the edge of a river demonstrate that tigers are not scared of water.

Fig. 14 Masks for protection. Masks on the back of their heads are supposed to protect them from a tiger attack when they go fishing in Sundarbans.

Whatever the reasons, the tigers of Sundarbans are behaving distinctively different than in all other areas of their distribution. People are therefore much more careful than anywhere else. No one, not even a ranger, would simply go by foot or on a bicycle in the marshes, as they normally do in all the other tiger reserves in the country. The villages located on the outskirts of the Sundarbans are surrounded by electric fences.
With the approval of the authorities, some people may enter the swamps in order to collect wood or honey. As a protection, they are often wearing masks on the back of the head depicting a second face. The belief is that tigers, who usually attack prey from behind, are confused and refrain from attacking (**fig. 14**). Puppets wearing human clothing have been charged-up with batteries so that »man-eaters« taking a bite are prevented from further attacks. For tourists visiting the Sundarbans special rules of behavior also apply. They may visit the area only accompanied by armed rangers and the access to the observation towers, where they are allowed to leave their boats, is always surrounded by tall, tiger secure fences.
Through all these measures, the number of deaths and attacks inflicted by tigers both in India and in Bangladesh have been greatly reduced. But conflicts are not over yet. Sundarbans has been and continues to be the area where man has to fear the tiger, as was initially described by Alfred Brehm.

Pl. 1 a-d Tigers are undoubtedly among the most beautiful and popular mammals, but also among the most feared. Different tigers in the Bandhavgarh National Park and in the Corbett National Park.

Pl. 1 (continued)

Pl. 2 **a** For wealthy hunters, also Indians, such a picture above the fireplace was and is an absolute hit. – **b** Tiger trap. Anyone who wants to defend himself against cattle hunters or kill a tiger for the illegal market makes use of such powerful slap traps. – **c** Degraded habitat. Even in the vicinity of national parks, here that of Ranthambore, the habitats are severely degraded by overgrazing. Kanha NP.

Pl. 3 **a** Tigers in front of riding elephants with tourists. Little impressed are tigers, which are visited in some reserves for one or two hours by riding elephants with tourists. – **b** Tigers with jeeps. Tigers also casually stroll past jeeps full of foreign and local tourists.

Pl. 4 **a** Axish deer. The lack of natural prey like *Axishirschen* could be a reason why in Sundarbans humans fall victim to the big cats. – **b** Wild boars. The Indian comb wild pigs belong to the most popular and most frequent booty animals of the tigers.

Pl. 5 **a** Sambar Deer. The mighty Sambar stags often become tiger victims, because they curiously approach their hunter only one step closer. – **b** Axis Stag. The dainty, but with a huge antler equipped *Axishirschen* is very careful and is therefore less often carried off.

Pl. 6 a-b Tigers in Thailand. Very much more rarely than the Indian king tigers are all other subspecies of the tigers, also the Indochina tigers held here by monks in Thailand.

Pl. 7 **a** Tiger statue in Bangladesh. Tiger statue behind fence: everywhere in the Sundarbans people protect rather themselves from the big cats in this way. – **b** This man was attacked by a tiger and only escaped with his life through the help of his friends.

Pl. 8 **a** The people living in the Sundarbans are very well aware of the danger posed by the big cats. – **b** An electric fence is to keep the dangerous tigers away from this village in the Sundarbans, but it is not always successful.

Pl. 9 **a** Fishermen in the Sundarbans on the border to Bangladesh. – **b** This tiger had killed a girl in the Sundarbans, was caught and brought to the Calcutta Zoo.

KLAUS MÜLLER-SCHILLING

BREEDING CONSERVATION OF ENDANGERED BIG CATS – THE EXAMPLE OF BARBARY LIONS

Zoological gardens allow the preservation of many endangered animal species. Modern zoos form arks for the conservation of these species, which are facing extinction due to environmental degradation.

WHAT IS A MODERN ZOO?

The world's oldest established zoo is Tiergarten Schönbrunn in Vienna, founded by the Austrian empress Maria Theresia. The golden age of zoo foundations was the 19th century, initiated by the middle classes. The animals were kept in cages. The most recent development in zoo concepts is as an animal theme park such as Erlebnis-Zoo Hannover, in which the visitor undertakes a journey on which he/she sees the animals in representations of their natural habitats.
A pioneer for modern zoos was the Swiss biologist and ethologist Heini Hediger (1908-1992). In 1942 he defined a new specialised branch of biology, zoo biology, in which he postulated that animals in zoos are not to be considered as »captives« but as »owners of property«, namely the territory of their enclosures. In modern zoos the quality of the enclosures (their furnishings and structure) is equally important for the captive animals. H. Hediger defined four main tasks for a modern zoo: recreation, education, research, and conservation. Today, modern zoos function as a present-day Noah's Ark. They have the experience, space and information (studbook) to save an endangered species. The IUCN (International Union for Conservation of Nature) states: »[…] more than 31 000 species are threatened with extinction. That is 27 % of all assessed species« (IUCN 2020). Thus modern zoos will be the last refuges of the endangered natural world.

HISTORY OF THE BARBARY LION IN NORTH AFRICA

Various authors (Yamaguchi/Haddane 2002; Hemmer/Burger 2005) have described in detail the history of the extinct Barbary lion (*Panthera leo leo*) and the status of the captive Moroccan »Royal Lion« collection as putative representatives of the subspecies. In 1826 the Austrian zoologist Johann Nepomuk Meyer was the first to describe Barbary lions as a subspecies *Felis leo barbaricus* (**fig. 1**) under the specific name *Felis leo* as used by Carl von Linné in 1758.

Fig. 1 Barbary lion in Algeria. – (Photo A. E. Pease, 1893).

In: Nicholas J. Conard · Henning Hassmann · Kurt Felix Hillgruber · Jordi Serangeli · Thomas Terberger (eds), The *Homotherium* Finds from Schöningen 13II-4. Man and Big Cats of the Ice Age. Forschungen zur Urgeschichte aus dem Tagebau von Schöningen 4 (Mainz 2022). DOI: https://doi.org/10.11588/propylaeum.1006.c13530

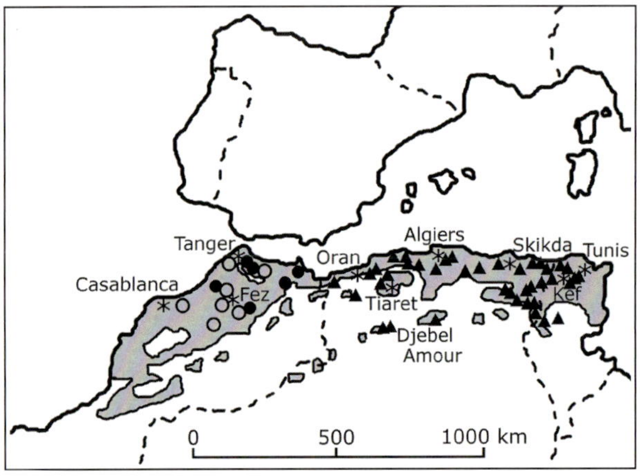

Fig. 2 Distribution of historical reports of lions in North Africa (1500-1900). – (After Black et al. 2013, 8 fig. 1).

The original habitat of the Barbary lion was North Africa, including mountainous regions from Morocco to Egypt (**fig. 2**). In the 18th century Barbary lions disappeared from Northeast Africa. In the mid-19th century the number of lions in Northwest Africa decreased (**fig. 3-4**). In 1942 the last recorded shooting of a Barbary lion, by a French colonial hunter, occurred in Morocco (Western Maghreb) (Yamaguchi/Haddane 2002; Black et al. 2013). Barbary lions were common in European menageries during the Middle Ages and more recently in public zoological parks (**figs 5-6**) and circuses up to their final extermination from the wild in the Atlas Mountains during the 1940s (IUCN 2020).

Since this time, some lion groups in captivity have been proposed as descendants of the North African Barbary lion. Especially the direct descendants from the King of Morocco's original collection, and their direct descendants in European zoos, have the strongest circumstantial claim (Yamaguchi/Haddane 2002).

Studies by Paul Leyhausen (1975) and Helmut Hemmer (1978) in the early 1970s identified the gene potential of lions in the Moroccan Royal collection, which they defined as »Royal Lions« at Rabat Zoo. These lions are presumably representatives of the current Barbary lion stock. A selective breeding programme involving all international zoos was proposed (Leyhausen 1975; Hemmer 1979; Frankham et al. 1986; Yamaguchi/Haddane 2002; Hemmer/Burger 2005).

The current Moroccan collection is descended from the individuals recorded in a census by Leyhausen (1975) and Hemmer (1978). Fortunately, all European individuals can be accurately traced to the 1974 set of Moroccan founders. In the European population a number of animals could be recorded as being descended from two individuals at Madrid Zoo, acquired from a circus in 1977 (ISIS 2008). Both were originally obtained from the royal palace collection prior to 1974. The Madrid lioness gave birth to cubs in July 1979, suggesting an age on arrival at Madrid of two years or more (Black et al. 2009).

Fig. 3 Lions at the Tower of London. – (Select views of London: with historical and descriptive sketches of some of the most interesting of the public buildings. Compiled and arranged by J. B. Papworth [London 1816]).

Fig. 4 Hunting of Barbary lions in the 19th century. – (agk-images, AKG4165093).

Fig. 5 Carte postale, Jardin des Plantes, Paris 1904.

REASONS FOR EXTINCTION

During the period of the Roman Empire Barbary lions fought for example against gladiators at the Roman Colosseum and other arenas; the Romans killed thousands of lions in their games (Pease 1899; Sparreboom 2016). In historical time, from the 16th century, the royal families of Morocco kept Bar-

Fig. 6 Another Barbary lion called Sultan, New York (Bronx) Zoo, 1897. – (After Nelson Robinson, Seventh Annual Report of the New York Zoological Society 1902 [1903], 121).

Fig. 7 Morphology of Barbary lions and Moroccan »Royal Lions« characteristics. – (Photo Erlebnis-Zoo Hannover, 2015).

bary lions. The Berbers offered lions in lieu of taxes and as gifts to the Sultan of Morocco and the Emperor of Ethopia, exacerbating their decline in the wild. During the colonial era lions were shot as coveted trophies, and the last Barbary lion was killed in Tunisia in 1891, in Algeria in 1893, and in 1942 the last in Morocco (Yamaguchi/Haddane 2002; Black et al. 2013).

Barbary lions were one of the biggest carnivores in Africa, hunting game such as Barbary sheep, Barbary stags and North African Hartebeest. These herbivores were also shot for trophies, depriving the lion of their prey in their habitat. Another problem was the strong interest of European and North American circuses and zoos in Barbary lions during the 19th and 20th centuries. Circuses and zoos were keen to present these powerful lions to their clientele and so many lions were captured and the wild population decreased even more. In the 1970s it was more common to give »Royal Lions« descendants directly to private collections. For example, between 1970 and 1973 Rabat Zoo sold 28 cubs to circuses and private buyers in Spain, Portugal and France (Yamaguchi/Haddane 2002). Thus, in these decades many of the »Royal Lions« offspring were transferred between various private institutions, specifically in North America and Europe (Hemmer 1979; Yamaguchi/Haddane 2002; Hemmer/Burger 2005).

LIVING IN CAPTIVITY

One of the places where lions were kept in captivity outside North Africa in the Middle Ages was the Tower of London Menagerie (**fig. 3**). This was stopped by an order of the Duke of Wellington in 1835 and the animals were transferred to London Zoo. One of the most popular London Zoo lions was called »Sultan« (1896). The Moroccan Sultans kept Barbary lions in historical times until 1912, when they were moved to the royal palace in Rabat. In the 1950s a few lions were transferred to Rabat Zoo, and by 1955 only 17 Barbary lions were kept at the royal palace. In 1973 the descendants of King Hassan II transferred all the lions to the Moroccan National Zoo in Rabat.

MORPHOLOGY AND BIOLOGY OF BARBARY LIONS

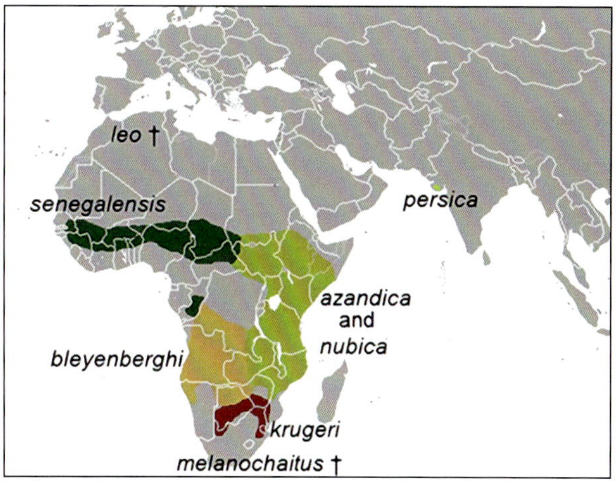

Fig. 8 Range map of the commonly accepted subspecies of the lion in the late 20th century. – (After Haas/Hayssen/Krausman 2005, 3).

Adult males weigh between 150 and 200kg, adult females weigh between 120 and 180kg. Lions range from 2.6 to 3.3m in length (including the tail); shoulder height is 1.2m. Barbary lions have a greyish pelage which, with their long fur, gives them a shaggy appearance. The females and young males have long hairs around the neck and throat, the back of the front legs and on the belly.

Adult males have a huge mane covering the head, neck, shoulders and belly (**fig. 7**). The colour of the mane varies, becoming darker toward the posterior parts. They have a well-developed tail tuft. They have a higher back of the head and a pointed crown, forming a straight line from the nose to back of the head, rounded cheeks and a narrow muzzle. These lions have a prominent anterior edge of the pelvis with dished rostral profile. The colour of the iris is egg-yolk yellow, not dark yellow or olive. Their food is a wide range of herbivorous prey including cattle, antelope and small game; like all lions they hunt in groups of lionesses in open and barren territory. Lions are a highly social species, an unusual characteristic amongst big

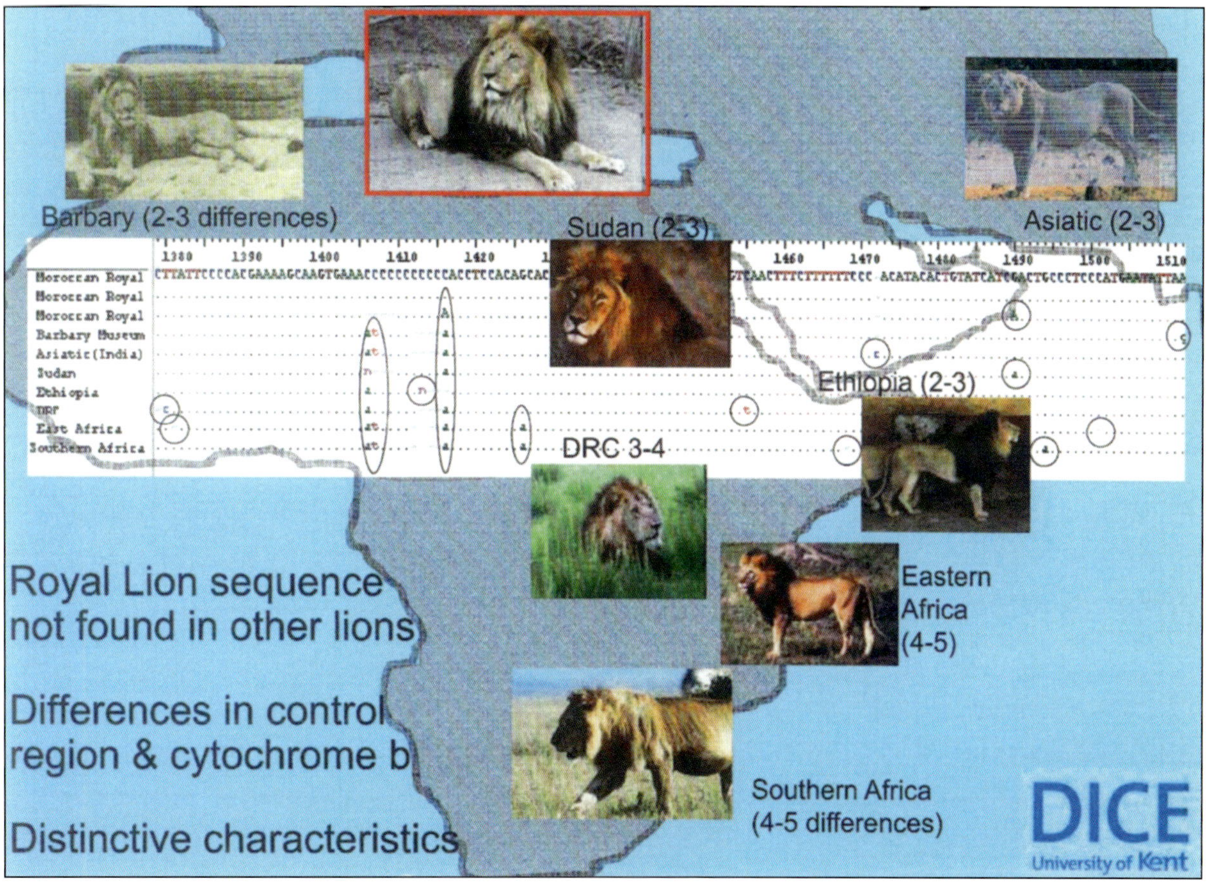

Fig. 9 Genetic analysis of Moroccan »Royal Lions«. – (After Harland/Black 2010).

cats, living in families and groups of 2-15 individuals. Females often give birth at the same time so the cubs can be reared together, suckling from any one of the group's mothers. Male cubs leave when they are sexually mature and roam together in bachelor groups. Lions have been divided into seven subspecies (**fig. 9**):

Genus: *Panthera* (Large cats)

- Species: *Panthera atrox* † (American lion)
- Species: *Panthera leo* (ion)
  - Subspecies: *Panthera leo bleyenberghi* (Southwest African or Katanga lion)
  - Subspecies: *Panthera leo goojratensis* (Asiatic lion)
  - Subspecies: *Panthera leo krugeri* (Southeast African or Transvaal lion)
  - Subspecies: *Panthera leo leo* (Barbary lion) †
  - Subspecies: *Panthera leo massaicus/nubia* (East African or Masai lion)
  - Subspecies: *Panthera leo melanochaita* (Cape lion) †
  - Subspecies: *Panthera leo persica* (Persian or Indian lion)
  - Subspecies: *Panthera leo senegalensis* (West African lion)
  - Subspecies: *Panthera leo azandica* (Congo lion)
  - Subspecies: *Panthera leo spelaea* (European cave lion) †

CONSERVATION IN CAPTIVITY

In 1970 there were only 35 purebred Barbary lions living at the Moroccan National Zoo in Rabat. In 1974 the National Zoo of Rabat asked German biologists Paul Leyhausen and Helmut Hemmer to study these lions to establish an international studbook. P. Leyhausen and H. Hemmer considered the »Royal Lions« only for their morphological and ethological traits and based on this studbook they established a Barbary lion survival project (Hemmer 1974; Burger/Hemmer 2006; Leyhausen 1975). The 1990s saw a revival of interest in »Royal Lions« at zoos of the European Association of Zoos and Aquaria (EAZA) and the Association of Zoos & Aquariums (AZA). At this time electronic records were collated from the handwritten breeding logbooks at the Rabat Zoo and other European and North American zoos that were keeping Barbary lions. Besides that, zoos were starting to use their data to inform other institutions for breeding decisions in conserving animals (Hill/Haynes 1999). By 1998 the maintenance of this data collecting programme had lapsed, and with it all clear inter-institutional commitment to preserving »Royal Lions« (Black et al. 2009).

Following studies from K. Hill and D. Haynes (1999) several European and American zoos launched the »Atlas Lion Project« and a selective breeding programme was proposed based on these results. However, there was no definite genetic relationship for the origin of the lions, and in 2005 Oxford University and Wildlinks International launched their »Atlas Lion Project«. For this project, sequences of mitochondrial DNA from 140-year-old museum samples were matched with the genetic material of living Barbary lions (**fig. 10**). Several zoos, which kept »Royal Lions«, abandoned their breeding programmes due to difficulties in obtaining new breeding stock, also being discouraged by the publication of genetic research which appeared to suggest that all lions shared a common ancestor (Yamaguchi 2000; Hemmer/Burger 2005). Consequently, there is a risk that the purity and genetic health of the bloodline may be compromised (compared to the original Moroccan royal collection **fig. 11**) in terms of inbreeding effects and a loss of genetic diversity. These problems would be exacerbated if future limited transfer of animals between institutions led to the isolation of breeding groups. However, several institutions remain committed to maintaining their »Royal Lion« collections, providing opportunities to manage the population appropriately (Black et al. 2009).

Fig. 10 Family tree for »Royal Lions« held in captive collections in Europe (1973-2009). – (After Harland/Black 2010).

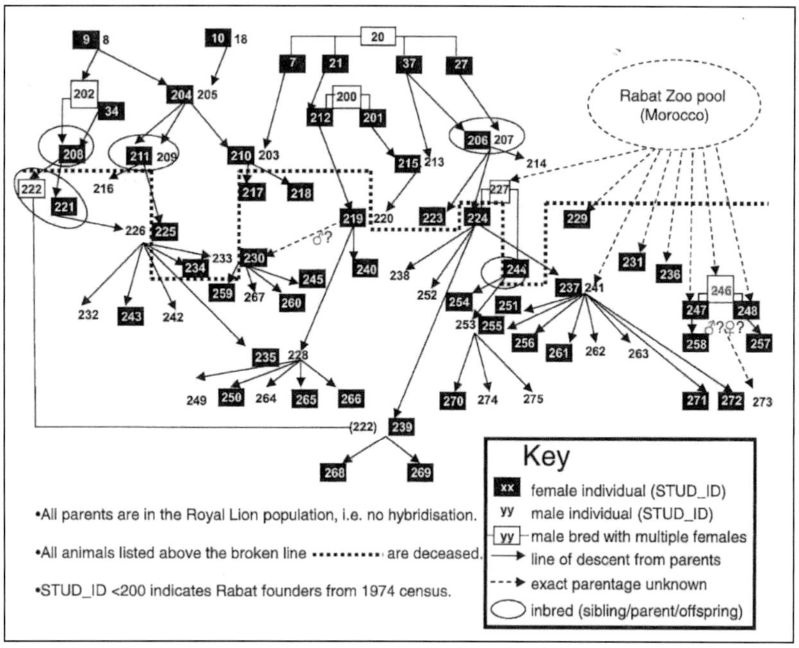

Fig. 11 Estimates of captive generations since wild collection in North Africa for current Moroccan Royal lions. – (S. A. Black / A. Fellous / N. Yamaguchi / D. L. Roberts, Examining the Extinction of the Barbary Lion and Its Implications for Felid Conservation. PLoS ONE 8(4): e60174 2013, fig. 4; https://doi.org/10.1371/journal.pone.0060174).

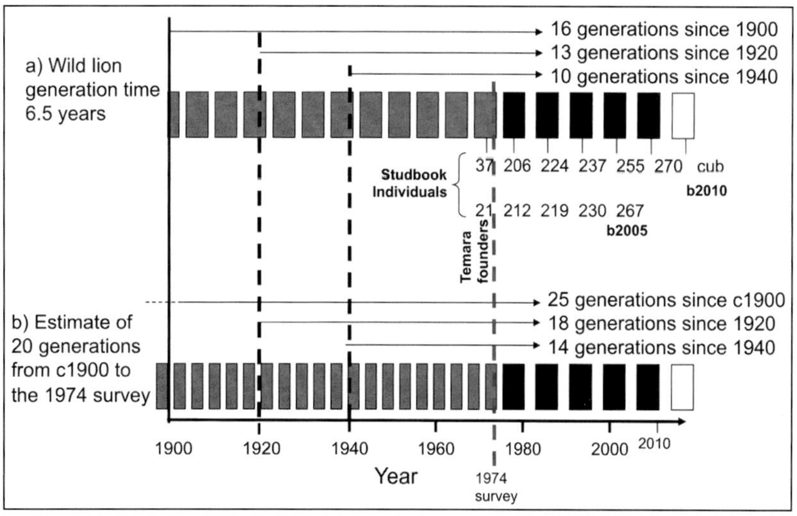

GENETIC RESULTS AND STUDBOOK

The genetic distinctiveness of the historical Barbary lion has not yet been fully established and the question of whether the »Royal Lions« are true Barbary lions also remains unanswered (Dubach et al. 2005; Yamaguchi 2005; Barnett et al. 2006b; Burger/Hemmer 2006; Antunes et al. 2008; Black et al. 2009). The historical Barbary lion is morphologically more distinct than any of the African lion populations (Hemmer 1979). As a consequence, guidelines were drawn up on the precautionary principle (Foster/Vecchia/Repacholi 2000). These guidelines would suggest that reasonable action to conserve diversity is preferable, with a separate management unit distinct from the other »zoo lions« in the maintenance of the »Moroccan Lions«. This study focuses on those European zoo collections (including Hai Kef Zoo in Israel), which hold reliable zoo records. In addition, Rabat Zoo holds approximately 25 individuals but breeding records are incomplete and do not appear on the International Species Information System (ISIS 2008). However, reasonable inferences about these animals can be made from historical records of selective breeding concerning the existing Rabat collection. The research team from Simon Black are reviewing the genetic ancestry of living animals and compiling a European studbook in order to identify (Black et al. 2009):

1. age and gender demographics;
2. reproductive success;
3. founder representation in European collections versus the Rabat group;
4. implications for a future breeding programme to maintain and improve the genetic health of this captive meta-population.

Information for supporting demographic assessment of the existing captive population was found in a number of primary and secondary sources, including the breeding records for »Royal Lions« since 1998. This data included information from both handwritten breeding records at the Rabat Zoo and the records in Western zoos (ISIS 2008). Recent records from zoos that include paternity and maternity data were published in official zoo publications and journal articles by L. Teichmann (2004) and L. Veselá et al. (2005). Informal sources included zoo websites, newsletters and personal communications between the authors and zoo staff (Black 2009). A printed version of the 1998 studbook for »Royal Lions« was validated against the ISIS species database (ISIS 2008). This information was analysed to identify founder animals (born before 1969) recorded in the H. Hemmer and P. Leyhausen census of the former royal palace collection (Leyhausen 1975), the breeding individuals from that population, and subsequent offspring and parentage for all other animals between 1970 and 1998 (Black et al. 2009).

In the studbook from 1998 S. Black and his team (2009) presented the current holdings with their parental relationships.

Current holdings comprised parental histories and individual animal entries for parents, using online ISIS queries, for comparison with the 1998 studbook (Black et al. 2009). Zoo websites and web-based zoo reports were scrutinized for supporting data on names, dates of birth, gender and parental identifiers. Several zoos claiming to hold Barbary lions were contacted directly by an email questionnaire. This process revealed that some institutions have either not submitted records to ISIS or have since withdrawn their claims to hold *Panthera leo leo* from the ISIS records (Black et al. 2009).

On the other hand the authors of the study wished to emphasise the Barbary lion label in marketing communications to the visitors. Data checking ensured the claim validity and the consistency of information cited in a number of different sources. Information on individual lions within the newly compiled studbook consists of zoo records going back to European collections included the known remaining animals from Hemmer and Leyhausen's 1974 census of pre-1969 founders (Black et al. 2009). The European collections included the known remaining animals from American zoos (Yamagucchi 2005) that participated in the original Hemmer and Leyhausen project (Leyhausen 1975; Hemmer 1979; Yamaguchi/Haddane 2002). Rabat Zoo does not have complete zoo records of maternity, paternity, birth date or other demographic information on their Barbary lion collection, so an equivalent studbook was not feasible. Each animal was allocated a unique reference number (STUD_ID) within the draft 2008 European studbook, because many individual animals have multiple zoo ID numbers and, in some instances, multiple names. Only living animals and their parental lines reaching back to the 1969 founder group were included in the studbook. Dead non-breeders or failed breeders (i.e. those with no surviving descendants) were excluded, although all living non-breeding animals have been included since genetic and morphological data for these animals may become important for future analyses. Analysis of parental lineage and founder gene representation involved identification of founder representation across the major zoo collections and then for the total known population of »Royal Lions« (i.e. zoos in Morocco, Europe and North America; **fig. 12**). The calculation involved mathematical proportioning of founder genes assuming an equal contribution from both sire and dam (Black et al. 2009).

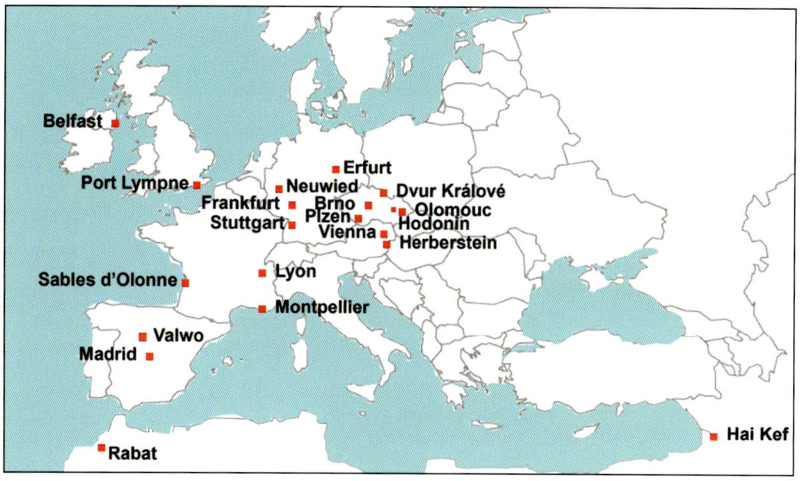

Fig. 12 Location of existing Moroccan »Royal Lions«. – (After Harland/Black 2010).

The spread of founder gene representation was calculated forwards from the royal palace founders with STUD_IDs 7-10, 18, 20-21, 27, 34, 37 and the two probable founders from Madrid Zoo. STUD_IDs 200, 201 (Yamaguchi/Haddane 2002; ISIS 2008) were the lions from Madrid with the ideal founder maximum of 8.33 %.

Based on these calculations, the proportion of founder genes represented in any one individual was estimated, together with the total proportion across the population. For Rabat Zoo lions, an estimate of founder representation was calculated as an average of the last known proportions of founder representation in 1978 (Black et al. 2009). This was adequate to generate calculations of founder gene representation in each subsequent set of offspring. The Microsoft Excel statistical functions enabled the analysis of means across sub-populations, based on the calculated founder representations for individual animals in each sub-population (Black et al. 2009). Demographic analysis of the »Royal Lion« population data from the draft 2008 European studbook was used to identify distributions of population age, gender, and opportunities for breeding pairs, cub mortality and fecundity and to reveal trends in these variables since the 1974 royal palace census (Black et al. 2009). At first samples of young females descended from the breeding group at Rabat Zoo and various other lion females were analysed for Cytochrome b sequences. The result of this first screening shows that the radiation within *Panthera leo* separating the extinct European cave lions from recent lions happened no more than 60 000 years ago (Antunes et al. 2008). Sub-Saharan and Asian lion sub-groups separated between 7400 and 20 000 years ago.

The genetic studies used mitochondrial DNA sequencing data to clarify the phylogenetic relationship between Barbary, Sub-Saharan and Asian lions. The phylogenetic tree (based on mtDNA) for the North African Barbary lion joins the Asian lion clade. It shows a slight difference but clear distinction from Sub-Saharan lions. It can be considered as a genetically defined phylogeographic group of its own. The date of splitting between Barbary lion and Asian lion happened later than the splitting of the other lions (Barnett et al. 2006a; 2006b; 2014). The results of the examination by Burger and Hemmer (2005) should be the basis of an urgent call for new strains for breeding of the stock of Barbary lions. Initial research into a possible future reintroduction into the Moroccan wilderness has recently been positively received by the Moroccan government.

The captive population of »Royal Lions« is currently held in relatively isolated zoo collections and is vulnerable to the effects of inbreeding depression (Balmford/Mace/Leader-Williams 1996). Pairing animals from UK zoos with those from Central Europe is a priority to increase diversity and retain a more even spread of founder genes. The draft 2008 studbook and the founder analysis in this study have identified a number of suitable breeding exchanges. A formal studbook-led breeding programme would enable constructive par-

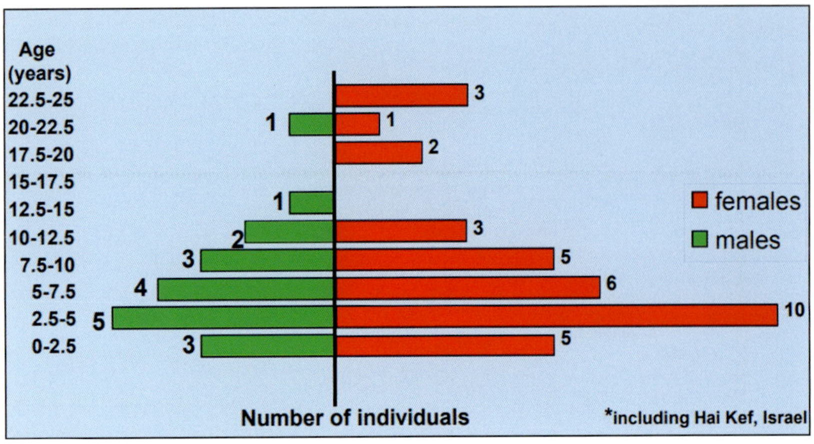

Fig. 13 Age profile for »Royal Lions«, differentiated by gender in European Zoos (including 1,1 lions at Hai Kef Zoo, Israel), Black, Yamaguchi, Harland, Groombridge 2009.

ticipation by zoos holding »Royal Lions« and allow future planning and negotiations (**fig. 13**). The Hemmer and Leyhausen morphological categorization has been largely ignored since the 1974 census (Leyhausen 1975; Hill/Haynes 1999; Yamaguchi/Haddane 2002; Tefera 2003) but should be incorporated into the studbook alongside genetic and parental data to enable easier assessment of pedigrees as new genetic knowledge of lions emerges (Dubach et al. 2005; Patterson 2007). Other zoo animals, if proven genetically similar to »Royal Lions«, could be included in future breeding efforts, whilst radical approaches such as in-vitro procedures might be considered to retain the genes of ageing »Royal Lions«. The holding capacity of participating zoos will be a constraint on population growth and institutions will need to be willing to transfer selected animals. However, opportunities exist for mutually beneficial exchanges of animals between zoo institutions that should enable effective maintenance of captive prides in the near future. The »Royal Lion« population may yet hold a unique genetic heritage for the global lion population (Patterson et al. 2005; Yamaguchi 2006) and the precautionary principle would suggest that reasonable steps should be taken to preserve »Royal Lions« until their conservation value has been properly assessed. The intention of the »Atlas Lion project group« comprising Simon Black, Nobuyuki Yamaguchi, Adrian Harland and Jim Groombridge (2009) was to create a studbook for the Barbary lion population in captivity.

All studbook data based on a review of zoo records in Rabat (1969-1998) plus a detailed review of breeding records from zoos worldwide (from 1974 onwards). The studbook could help return the majestic Barbary lion to the wild, and will help establish a breeding programme for the Barbary lion. Black's team managed and coordinated breeding to optimize the overall captive population of Moroccan »Royal Lions«. At the moment Barbary lions are managed by EAZA monitoring. Currently we have 54 living individuals in the European zoo population (**fig. 13**), with 15 males, 31 females, 8 juveniles (3 ♂ and 5 ♀).

HANOVER ZOO BARBARY LIONS

Hanover Zoo opened in 1865, founded on the initiative of the Hanoverian middle class. From 1931 till 1971 the animal dealer Herman Ruhe was the owner of Hanover Zoo, and from 1972 till 1993 the zoo was owned and managed by Hanover city council. Since 1993 the zoo has been a limited liability company under the supervision of the Kommunalverband Großraum Hannover.

As an EXPO 2000 World Exposition project, Hanover Zoo began to transform and modernise its grounds, replacing cages with spacious enclosures comprising theme worlds. One of their main focuses is the nature

Fig. 14 View from the visitors' platform at Hanover Zoo. – (Photo Erlebnis-Zoo Hannover, 2015).

and endangered species of North Africa. Thus Hanover Zoo hold the studbooks for Addax and North African ostriches. We keep, among others, endangered species such as Somali wild ass and since 2010, the extinct lion subspecies of Barbary lions (**fig. 15**). The lion enclosure was built in 2000.

At first we kept only three »zoo lions« from Givskud Zoo in Denmark. The enclosure is built to resemble an African canyon. The outdoor facility covers 520 m², bounded at the front by an 8 m wide and 5 m deep dry ditch. Visitors can see into the enclosure through four windows.

From the big window one has a fine view across the lion enclosure to our ungulates enclosure with Hartman's mountain zebras, blesboks, springboks and Rothschild's giraffes. The lions have two high lookout points as important features in the enclosure (**fig. 16**). The artificial rock boundary wall is 5 m high. The indoor facility covers 41 m².

THE BARBARY LIONS AT ERLEBNIS-ZOO HANOVER

The adult Hanoverian lion male Chalid was captive born on the 15th of April 2005 at Port Lympne / UK, and is now 10 years old (**fig. 17a**). Chalid has lived at the Hanover Zoo since the 30th of June 2010. He sired five offspring so far. In 2011 couple Barbary lions were born. The young male Joco and his sister Zari were the first Barbary lions to be born in Hanover. Three years later Chalid fathered three more lions. The male, Basu, and his sisters Tamika and Neyla were born on the 1st of July 2015.

Fig. 15 View of the lion enclosure at Hanover Zoo. – (Photo Erlebnis-Zoo Hannover 2015).

Fig. 16 View from the visitor platform across the boundary ditch into the lion enclosure at Hanover Zoo. – (Photo Erlebnis-Zoo Hannover 2015).

Fig. 17 Barbary lions in Erlebnis-Zoo Hannover: **a** Chalid. – **b** Binta with cubs. – **c** Naima. – **d** lion cubs in 2011. – **e** lion cubs in 2014. – (Photos Erlebnis-Zoo Hannover).

Fig. 17 (continued)

Fig. 17 (continued)

The mother of these lion cubs is the Barbary lioness Binta, one of the grandchildren of the »Moroccan Palace Barbary Lions«. Binta is 7 years old and has lived at Hanover Zoo since the 1st of March 2010 (**fig. 17b**). The other lioness, Naima, was also captive-born at Rabat Zoo in Morocco (**fig. 17c**). She is 7 years old and has lived at Hannover Zoo since the 1st of March 2010. Naima is Binta's sister and has no offspring; because of health problems in their uterine tube, we have been treating Naima with hormone contraception since 2010.

The five cubs were born in 2011 and 2014, two males and three females. Both firtborn Barbary lions now live at Thüringer Zoopark Erfurt (Joco) and the female Zari was sent to the Zoo Neuwied (**fig. 19d**).

Of the three litter the females Tamika and Neyla now live at Plzeň Zoo in the Czech Republic. The young male Basu has so far stayed in Hanover (**fig. 17e**). We hope that all Hanover-born animals will do their share for the conservation of the species, extinct in the wild.

REFERENCES

Antunes et al. 2008: A. Antunes / J. L. Toyer / M. E. Roelke / J. Pecon-Slattery / C. Packer / C. Winterbach / H. Winterbach / G. Hemson / L. Frank / P. Stander / L. Siefert / M. Driciru / P. J. Funston / K. A. Alexander / K. C. Prager / G. Mills / D. Wildt / M. Bush / S. J. O'Brien / W. E. Johnson, The Evolutionary Dynamics of the Lion *Panthera leo* Revealed by Host and Viral Population Genomics. PLoS Genet 4/11, 2008, e1000251. DOI: 10.1371/journal.pgen.1000251.

Balmford/Mace/Leader-Williams 1996: A. Blamford / G. M. Mace / N. Leader-Williams, Designing the Ark: Setting Priorities for Captive Breeding. Conservation Biology 10/3, 1996, 719-727.

Barnett et al. 2006a: R. Barnett / N. Yamaguchi / I. Barnes / A. Cooper, The Origin, Current Diversity and Future Conservation of the Modern Lion (*Panthera leo*). Proceedings of the Royal Society: Biological Sciences 273/1598, 2006, 2119-2125.

2006b: R. Barnett / N. Yamaguchi / I. Barnes / A. Cooper, Lost Populations and Preserving Genetic Diversity in the Lion *Panthera leo*: Implications for its Ex Situ Conservation. Conservation Genetics 7, 2006, 507-514.

2008: R. Barnett / N. Yamaguchi / B. Shapiro / R. A. Sabin, Ancient DNA Analysis Indicates the First English Lions Originated from North Africa. Contributions to Zoology 77/1, 2008, 7-16.

2014: R. Barnett / N. Yamaguchi / B. Shapiro / S. Y. W. Ho / I. Barnes / R. Sabin / L. Werdelin / J. Cuisin / G. Larson, Revealing the Maternal Demographic History of *Panthera leo* Using Ancient DNA and a Spatially Explicit Genealogical Analysis. BMC Evolutionary Biology 14/70, 2014. DOI: 10.1186/1471-2148-14-70.

Black et al. 2009: S. A. Black / N. Yamaguchi / A. Harland / J. Groombridge, Maintaining the Genetic Health of Putative Barbary lions in Captivity: an Analysis of Moroccan Royal Lions. European Journal of Wildlife Research 56/1, 2009, 21-31.

2013: S. A. Black / A. Fellous / N. Yamaguchi / D. L. Roberts, Examining the Extinction of the Barbary Lion and Its Implications for Felid Conservation. PLoS ONE 8/4, 2013, e60174. DOI: 10.1371/journal.pone.0060174.

Burger/Hemmer 2006: J. Burger / H. Hemmer, Urgent Call for Further Breeding of the Relic Zoo Population of the Critically Endangered Barbary lion (*Panthera leo leo* Linnaeus 1758). European Journal of Wildlife Research 52, 2006, 54-58.

Dubach et al. 2005: J. Dubach / B. D. Patterson / M. B. Biggs / K. Venzke / J. Flamand / P. Stander / L. Scheepers / R. W. Kays, Molecular Genetic Variation Across Southern and Eastern Geographic Ranges of the African Lion, *Panthera leo*. Conservation Genetics 6, 2005, 15-24.

Foster/Vecchia/Repacholi 2000: K. R. Foster / P. Vecchia / M. H. Repacholi, Science and the precautionary principle. Science 288/5468, 2000, 979-981.

Frankham et al. 1986: R. Frankham / H. Hemmer / O. A. Ryder / E. G. Cothran / M. E. Soulé / N. D. Murray / M. Snyder, Selection in Captive Populations. Zoo Biology 5/2, 1986, 127-138.

Geiser 1875: J. Geiser, Lion de L'atlas (Alger 1875).

Harland/Black 2010: A. Harland / S. Black, Moroccan Royal Lions: justification for a new EEP to support lion sub-species conservation [Felid TAG Meeting presentation].

Haas/Hayssen/Krausman 2005: S. K. Haas / V. Hayssen / P. R. Krausman, *Panthera leo*. American Society of Mammalogists 762, 2005, 1-11.

Hemmer 1967: H. Hemmer, Fossilbelege zur Verbreitung und Artgeschichte des Löwen, *Panthera leo* (Linné, 1758). Säugetierkundliche Mitteilungen 15, 1967, 289-300.

1974: H. Hemmer, Untersuchungen zur Stammesgeschichte der Pantherkatzen (Pantherinae), Teil III: Zur Artgeschichte des Löwen *Panthera* (*Panthera*) *leo* (Linnaeus 1758). Veröffentlichungen der Zoologischen Staatssammlung München 17, 1974, 167-280.

1979: H. Hemmer, Grundlagen und derzeitiger Stand des Zuchtprogrammes zur Rückerhaltung der Berberlöwen (*Panthera leo leo*). In: S. Seifert / P. Müller (eds), Kongressbericht. 1. Internationales Symposium über die Haltung und Zucht des Tigers. 11. und 12. Oktober 1978 in Leipzig (Leipzig 1979) 65-72.

Hill/Haynes 1999: K. Hill / D. Haynes, Atlas lion preservation scheme project description (Port Lympne). International Zoo News 52, 1999, 218-223.

IUCN 2020: The IUCN Red List of Threatened Species. http://iucnredlist.org.

Leyhausen 1975: P. Leyhausen, Preliminary Report on the Possibility of a Breeding Programme for the Atlas Lion at Temara Zoo, Morocco. International Zoo News 21, 1975, 22-23.

O'Brien et al. 1987: S. O'Brien / J. S. Martenson / C. Parker / L. Herbst / V. de Vos / P. Joslin / D. E. Wildt / M. Bush, Biochemical Genetic Variation in Geographic Isolates of African and Asiatic Lions. National Geographic Research 3, 1987, 114-124.

Patterson 2007: B. D. Patterson, On the Nature and Significance of Variability in Lions (*Panthera leo*). Evolutionary Biology 34, 2007, 55-60.

Patterson et al. 2005: B. D. Patterson / N. Yamaguchi / J. M. Dubach / D. York, Molecular Genetics and Morphological Variation of Lions (*Panthera leo*). African Lion News 6, 2005, 17-23.

Pease 1899: A. E. Pease, The lion in Algeria. In: A. J. Arnold / H. A. Bryden / T. E. Buckley (eds), Great and small game of Africa. An account of the distribution, habits and natural history of the sporting mammals, with personal hunting experiences (London 1899) 564-568.

1913: A. E. Pease, The Distribution of Lions. In: A. E. Pease, The Book of the Lion (London 1913) 109-147.

Seifert 1978: S. Seifert, Untersuchungen zur Fortpflanzungsbiologie der im Zoologischen Garten Leipzig gehaltenen Großkatzen (*Panthera*, Oken, 1816) unter besonderer Berücksichtigung des Löwen-*Panthera leo* (Linné, 1758) (Berlin 1978).

Sparreboom 2016: A. Sparreboom, Chapter 2: Procuring beasts for hunting spectacles. In: A. Sparreboom, Venationes Africanae: Hunting spectacles in Roman North Africa: cultural significance and social function (Amsterdam 2016) 67-98.

Tefera 2003: M. Tefera, Phenotypic and Reproductive Characteristics of Lions (*Panthera leo*) at Addis Ababa Zoo. Biodiversity and Conservation 12, 2003, 1629-1639.

Teichmann 2004: L. Teichmann, Berberlöwen im Zoo Neuwied. Gemeinschaft Deutscher Zooförderer e. V. Almanach 6, 2003/2004 (2004), 30.

Veselá et al. 2005: L. Veselá / M. Kofiínek / J. Vokurková / H. Labská, Chov lvů berberských v ZOO Olomouc. Breeding of Barbary lion in Zoo Olomouc. Zoologická zahrada. Zoological Garden Olomouc, Czech Republic, Zoo Olomouc – Výroční Zpráva. The Annual Report 2005, 35-39.

Yamaguchi 2000: N. Yamaguchi, Follow up: the Barbary lion Project: its feasibility and potential. African Lion News 2, 2000, 12-13.

2005: N. Yamaguchi, The identity of the Royal lion of Morocco: a recent development in molecular investigation. African Lion News 6, 2005, 23-26.

2006: N. Yamaguchi, Re-evaluation of »generic« zoo lions. International Zoo News 53, 2006, 41-43.

Yamaguchi/Haddane 2002: N. Yamaguchi / B. Haddane, The North African Barbary lion and the Atlas lion Project. International Zoo News 49/8, 2002, 465-481.

SUMMARY / ZUSAMMENFASSUNG

Breeding Conservation of Endangered Big Cats – The Example of Barbary Lions
Barbary lions have been extinct in the wild since 1942. There is a promising zoo population of 54 animals: 15 males, 31 females and 8 juveniles (3♂, 5♀). Modern zoos are acting as a Noah's Ark – they have experience, space and information (studbook, genetic analysis). There is a chance to reintroduce this species in the wild. However, zoos are closed spaces with all variety of comforts, such as daily feeding, cleaning the facilities and medical care. Thinking about returning to the wild has to be part of the breeding strategy. In the case of the Barbary lion, we are hopeful that it will be possible to eventually return them to the wild.

Zuchterhaltung gefährdeter Großkatzen – das Beispiel der Berberlöwen
Berberlöwen sind in freier Wildbahn seit 1942 ausgerottet. Es gibt eine vielversprechende Zoopopulation von 54 Tieren: 15 Männchen, 31 Weibchen und 8 Jungtiere (3♂, 5♀). Moderne Zoos fungieren als Arche Noah – sie haben Erfahrung, Platz und Informationen (Zuchtbuch, genetische Analysen). Es besteht die Chance, diese Art wieder in der freien Natur anzusiedeln. Allerdings sind Zoos geschlossene Räume mit allen möglichen Annehmlichkeiten wie täglicher Fütterung, Reinigung der Anlagen und medizinischer Versorgung. Der Gedanke an eine Rückkehr in die freie Wildbahn muss Teil der Zuchtstrategie sein. Im Falle des Berberlöwen sind wir zuversichtlich, dass es möglich sein wird, ihn in die freie Wildbahn zu entlassen.

LIST OF CONTRIBUTORS

Mauricio Antón
Museo Nacional de Ciencias Naturales
Departamento de Paleobiología
C. José Gutiérrez Abascal, 2
E - 28006 Madrid
mauricioanton24@gmail.com

Ross Barnett
5 Bellfield Terrace
UK - Inverness, Scotland, IV24ST
drrossbarnett@gmail.com

Kurt Felix Hillgruber
Archäologisches Landesmuseum Baden-Württemberg
Benediktinerplatz 5
D - 78467 Konstanz
hillgruber@konstanz.alm-bw.de

Fritz Jantschke
fritz.jantschke@web.de

Ulrich Joger
Staatliches Naturhistorisches Museum
Pockelsstr. 10
D - 38106 Braunschweig
u.joger@3landesmuseen.de

Thijs van Kolfschoten
Universiteit Leiden
Faculteit der Archeologie
Einsteinweg 2
NL - 2333 CC Leiden
t.van.kolfschoten@arch.leidenuniv.nl
and
Shandong University
Institute of Cultural Heritage
72 Binhai Highway
CN - 266237 Qingdao

Bram W. Langeveld
Natuurhistorisch Museum Rotterdam
Westzeedijk 345
NL - 3015 AA Rotterdam
langeveld@hetnatuurhistorisch.nl

Dick Mol
Natuurhistorisch Museum Rotterdam
Westzeedijk 345
NL - 3015 AA Rotterdam
dickmol@telfort.nl

Klaus Müller-Schilling
Zoo Hannover gGmbH
Adenauerallee 3
D - 30175 Hannover
klaus.mueller-schilling@erlebnis-zoo.de

Elena A. Nikulina
Zentrum für Baltische und Skandinavische Archäologie
(ZBSA)
Stiftung Schleswig-Holsteinische Landesmuseen
Schloss Gottorf
D - 24837 Schleswig
elena.nikulina@schloss-gottorf.de

André Ramcharan
Universiteit Leiden
Faculteit der Archeologie
Einsteinweg 2
NL - 2333 CC Leiden
a.ramcharan@arch.leidenuniv.nl

Ulrich Schmölcke
Zentrum für Baltische und Skandinavische Archäologie
(ZBSA)
Stiftung Schleswig-Holsteinische Landesmuseen
Schloss Gottorf
D - 24837 Schleswig
ulrich.schmoelcke@schloss-gottorf.de

Jordi Serangeli
Senckenberg Centre for Human Evolution and
Palaeoenvironment an der Universität Tübingen
palaeon 1
D - 38364 Schöningen
jordi.serangeli@uni-tuebingen.de

Britt Starkovich
Eberhard Karls Universität Tübingen
Urgeschichte und Naturwissenschaftliche Archäologie
Senckenberg Centre for Human Evolution and
Palaeoenvironment an der Universität Tübingen
Rümelinstr. 23
D - 72070 Tübingen
britt.starkovich@uni-tuebingen.de

Clara Stefen
Senckenberg Naturhistorische Sammlungen Dresden
Museum für Tierkunde
Königsbrücker Landstr. 159
D - 01127 Dresden
clara.stefen@senckenberg.de

Anthony J. Stuart
Durham University
Biological and Biomedical Sciences
South Road
UK - Durham DH1 3LE

Thomas Terberger
Niedersächsisches Landesamt für Denkmalpflege
Scharnhorststr. 1
D - 30175 Hannover
thomas.terberger@nld.niedersachsen.de

Ivo Verheijen
Senckenberg Centre for Human Evolution and
Palaeoenvironment an der Universität Tübingen
palaeon 1
D - 38364 Schöningen
theskullcollector@hotmail.com
and
Universiteit Leiden
Faculteit der Archeologie
Einsteinweg 2
NL - 2333 CC Leiden

Aritza Villaluenga
Universidad del País Vasco-Euskal Herriko Unibertsitatea
(UPV-EHU)
Facultad de Letras, Edificio Anejo. Lab. 1.19A
C/ Francisco Tomás y Valiente s/n
E - 01006 Vitoria-Gasteiz, Álava-Araba

Nicolas Zenzen
Henneberg-Museum
Deutschherrnstr. 18
D - 97702 Münnerstadt
nzenzen@gmx.de